An Introduction to Ramsey Theory

STUDENT MATHEMATICAL LIBRARY
Volume 87

An Introduction to Ramsey Theory

Fast Functions, Infinity, and Metamathematics

Matthew Katz

Jan Reimann

Editorial Board

Satyan L. Devadoss
Rosa Orellana

John Stillwell (Chair)
Serge Tabachnikov

2010 *Mathematics Subject Classification*. Primary 05D10, 03-01, 03E10, 03B10, 03B25, 03D20, 03H15.

Jan Reimann was partially supported by NSF Grant DMS-1201263.

For additional information and updates on this book, visit
www.ams.org/bookpages/stml-87

Library of Congress Cataloging-in-Publication Data
Names: Katz, Matthew, 1986– author. | Reimann, Jan, 1971– author. | Pennsylvania State University. Mathematics Advanced Study Semesters.
Title: An introduction to Ramsey theory: Fast functions, infinity, and metamathematics / Matthew Katz, Jan Reimann.
Description: Providence, Rhode Island: American Mathematical Society, [2018] | Series: Student mathematical library; 87 | "Mathematics Advanced Study Semesters." | Includes bibliographical references and index.
Identifiers: LCCN 2018024651 | ISBN 9781470442903 (alk. paper)
Subjects: LCSH: Ramsey theory. | Combinatorial analysis. | AMS: Combinatorics – Extremal combinatorics – Ramsey theory. msc | Mathematical logic and foundations – Instructional exposition (textbooks, tutorial papers, etc.). msc | Mathematical logic and foundations – Set theory – Ordinal and cardinal numbers. msc | Mathematical logic and foundations – General logic – Classical first-order logic. msc | Mathematical logic and foundations – General logic – Decidability of theories and sets of sentences. msc | Mathematical logic and foundations – Computability and recursion theory – Recursive functions and relations, subrecursive hierarchies. msc | Mathematical logic and foundations – Nonstandard models – Nonstandard models of arithmetic. msc
Classification: LCC QA165 .K38 2018 | DDC 511/.66–dc23
LC record available at https://lccn.loc.gov/2018024651

Copying and reprinting. Individual readers of this publication, and nonprofit libraries acting for them, are permitted to make fair use of the material, such as to copy select pages for use in teaching or research. Permission is granted to quote brief passages from this publication in reviews, provided the customary acknowledgment of the source is given.

Republication, systematic copying, or multiple reproduction of any material in this publication is permitted only under license from the American Mathematical Society. Requests for permission to reuse portions of AMS publication content are handled by the Copyright Clearance Center. For more information, please visit www.ams.org/publications/pubpermissions.

Send requests for translation rights and licensed reprints to reprint-permission @ams.org.

© 2018 by the authors. All rights reserved.
Printed in the United States of America.

∞ The paper used in this book is acid-free and falls within the guidelines established to ensure permanence and durability.
Visit the AMS home page at https://www.ams.org/

10 9 8 7 6 5 4 3 2 1   23 22 21 20 19 18

Contents

Foreword: MASS at Penn State University   vii

Preface   ix

Chapter 1. Graph Ramsey theory   1
§1.1. The basic setting   1
§1.2. The basics of graph theory   4
§1.3. Ramsey's theorem for graphs   14
§1.4. Ramsey numbers and the probabilistic method   21
§1.5. Turán's theorem   31
§1.6. The finite Ramsey theorem   34

Chapter 2. Infinite Ramsey theory   41
§2.1. The infinite Ramsey theorem   41
§2.2. König's lemma and compactness   43
§2.3. Some topology   50
§2.4. Ordinals, well-orderings, and the axiom of choice   55
§2.5. Cardinality and cardinal numbers   64
§2.6. Ramsey theorems for uncountable cardinals   70
§2.7. Large cardinals and Ramsey cardinals   80

Chapter 3.	Growth of Ramsey functions	85
§3.1.	Van der Waerden's theorem	85
§3.2.	Growth of van der Waerden bounds	98
§3.3.	Hierarchies of growth	105
§3.4.	The Hales-Jewett theorem	113
§3.5.	A really fast-growing Ramsey function	123
Chapter 4.	Metamathematics	129
§4.1.	Proof and truth	129
§4.2.	Non-standard models of Peano arithmetic	145
§4.3.	Ramsey theory in Peano arithmetic	152
§4.4.	Incompleteness	159
§4.5.	Indiscernibles	171
§4.6.	Diagonal indiscernibles via Ramsey theory	182
§4.7.	The Paris-Harrington theorem	188
§4.8.	More incompleteness	193
Bibliography		199
Notation		203
Index		205

Foreword: MASS at Penn State University

This book is part of a collection published jointly by the American Mathematical Society and the MASS (Mathematics Advanced Study Semesters) program as a part of the Student Mathematical Library series. The books in the collection are based on lecture notes for advanced undergraduate topics courses taught at the MASS (Mathematics Advanced Study Semesters) program at Penn State. Each book presents a self-contained exposition of a non-standard mathematical topic, often related to current research areas, which is accessible to undergraduate students familiar with an equivalent of two years of standard college mathematics, and is suitable as a text for an upper division undergraduate course.

Started in 1996, MASS is a semester-long program for advanced undergraduate students from across the USA. The program's curriculum amounts to sixteen credit hours. It includes three core courses from the general areas of algebra/number theory, geometry/topology, and analysis/dynamical systems, custom designed every year; an interdisciplinary seminar; and a special colloquium. In addition, every participant completes three research projects, one for each core course. The participants are fully immersed into mathematics, and this, as well as intensive interaction among the students, usually leads to a dramatic increase in their mathematical enthusiasm and achievement. The program is unique for its kind in the United States.

Detailed information about the MASS program at Penn State can be found on the website `www.math.psu.edu/mass`.

Preface

If we split a set into two parts, will at least one of the parts behave like the whole? Certainly not in every aspect. But if we are interested only in the persistence of *certain small regular substructures*, the answer turns out to be "yes".

A famous example is the persistence of *arithmetic progressions*. The numbers $1, 2, \ldots, N$ form the most simple arithmetic progression imaginable: The next number differs from the previous one by exactly 1. But the numbers $4, 7, 10, 13, \ldots$ also form an arithmetic progression, where each number differs from its predecessor by 3.

So, if we split the set $\{1, \ldots, N\}$ into two parts, will one of them contain an arithmetic progression, say of length 7? *Van der Waerden's theorem*, one of the central results of Ramsey theory, tells us precisely that: *For every k there exists a number N such that if we split the set $\{1, \ldots, N\}$ into two parts, one of the parts contains an arithmetic progression of length k.*

Van der Waerden's theorem exhibits the two phenomena, the interplay of which is at the heart of Ramsey theory:

- **Principle 1:** If we split a large enough object with a certain regularity property (such as a set containing a long arithmetic progression) into two parts, one of the parts will also exhibit this property (to a certain degree).

- **Principle 2:** When proving Principle 1, "large enough" often means *very, very, very large*.

The largeness of the numbers encountered seems intrinsic to Ramsey theory and is one of its most peculiar and challenging features. Many great results in Ramsey theory are actually new proofs of known results, but the new proofs yield much better bounds on *how large* an object has to be in order for a Ramsey-type persistence under partitions to take place. Sometimes, "large enough" is even so large that the numbers become difficult to describe using axiomatic arithmetic—so large that they venture into the realm of *metamathematics*.

One of the central issues of metamathematics is *provability*. Suppose we have a set of *axioms*, such as the group axioms or the axioms for a vector space. When you open a textbook on group theory or linear algebra, you will find results (theorems) that follow from these axioms by means of logical deduction. But how does one know whether a certain statement about groups is provable (or refutable) from the axioms at all? A famous instance of this problem is Euclid's fifth postulate (axiom), also known as the *parallel postulate*. For more than two thousand years, mathematicians tried to derive the parallel postulate from the first four postulates. In the 19th century it was finally discovered that the parallel postulate is *independent* of the first four axioms, that is, neither the postulate nor its negation is entailed by the first four postulates.

Toward the end of the 19th century, mathematicians became increasingly disturbed as more and more strange and paradoxical results appeared. There were different sizes of infinity, one-dimensional curves that completely fill two-dimensional regions, and subsets of the real number line that have no reasonable measure of length, or there was the paradox of a set containing all sets not containing themselves. It seemed increasingly important to lay a solid foundation for mathematics. David Hilbert was one of the foremost leaders of this movement. He suggested finding axiom systems from which all of mathematics could be formally derived and in which it would be impossible to derive any logical inconsistencies.

An important part of any such foundation would be axioms which describe the natural numbers and the basic operations we perform on

Preface

them, addition and multiplication. In 1931, Kurt Gödel published his famous *incompleteness theorems*, which dealt a severe blow to Hilbert's program: For any reasonable, consistent axiomatization of arithmetic, there are independent statements—statements which can be neither proved nor refuted from the axioms.

The independent statements that Gödel's proof produces, however, are of a rather artificial nature. In 1977, Paris and Harrington found a result in Ramsey theory that is independent of arithmetic. In fact, their theorem is a seemingly small variation of the original Ramsey theorem. It is precisely the *very rapid growth of the Ramsey numbers* (recall Principle 2 above) associated with this variation of Ramsey's theorem that makes the theorem unprovable in Peano arithmetic.

But if the Paris-Harrington principle is unprovable in arithmetic, how do we convince ourselves that it is true? We have to pass from the finite to the *infinite*. Van der Waerden's theorem above is of a finitary nature: All sets, objects, and numbers involved are finite. However, basic Ramsey phenomena also manifest themselves when we look at infinite sets, graphs, and so on. Infinite Ramsey theorems in turn can be used (and, as the result by Paris and Harrington shows, sometimes have to be used) to deduce finite versions using the *compactness principle*, a special instance of topological compactness. If we are considering only the infinite as opposed to the finite, Principle 2 in many cases no longer applies.

- **Principle 1 (infinite version):** If we split an infinite object with a certain regularity property (such as a set containing arbitrarily long arithmetic progressions) into two parts, one infinite part will exhibit this property, too.

If we take into account, on the other hand, that there are different sizes of infinity, as reflected by Cantor's theory of ordinals and cardinals, Principle 2 reappears in a very interesting way. Moreover, as with the Paris-Harrington theorem, it leads to metamathematical issues, this time in *set theory*.

It is the main goal of this book to introduce the reader to the interplay between Principles 1 and 2, from finite combinatorics to set theory to metamathematics. The book is structured as follows.

In Chapter 1, we prove Ramsey's theorem and study Ramsey numbers and how large they can be. We will make use of the probabilistic methods of Paul Erdős to give lower bounds for the Ramsey numbers and a result in extremal graph theory.

In Chapter 2, we prove an infinite version of Ramsey's theorem and describe how theorems about infinite sets can be used to prove theorems about finite sets via compactness arguments. We will use such a strategy to give a new proof of Ramsey's theorem. We also connect these arguments to topological compactness. We introduce ordinal and cardinal numbers and consider generalizations of Ramsey's theorem to uncountable cardinals.

Chapter 3 investigates other classical Ramsey-type problems and the large numbers involved. We will encounter fast-growing functions and make an analysis of these in the context of primitive recursive functions and the Grzegorczyk hierarchy. Shelah's elegant proof of the Hales-Jewett theorem, and a Ramsey-type theorem with truly explosive bounds due to Paris and Harrington, close out the chapter.

Chapter 4 deals with metamathematical aspects. We introduce basic concepts of mathematical logic such as proof and truth, and we discuss Gödel's completeness and incompleteness theorems. A large part of the chapter is dedicated to formulating and proving the Paris-Harrington theorem.

The results covered in this book are all cornerstones of Ramsey theory, but they represent only a small fraction of this fast-growing field. Many important results are only briefly mentioned or not addressed at all. The same applies to important developments such as ultrafilters, structural Ramsey theory, and the connection with dynamical systems. This is done in favor of providing a more complete narrative explaining and connecting the results.

The unsurpassed classic on Ramsey theory by Graham, Rothschild, and Spencer [24] covers a tremendous variety of results. For those especially interested in Ramsey theory on the integers, the book

Preface

by Landman and Robertson [**43**] is a rich source. Other reading suggestions are given throughout the text.

The text should be accessible to anyone who has completed a first set of proof-based math courses, such as abstract algebra and analysis. In particular, no prior knowledge of mathematical logic is required. The material is therefore presented rather informally at times, especially in Chapters 2 and 4. The reader may wish to consult a textbook on logic, such as the books by Enderton [**13**] and Rautenberg [**54**], from time to time for more details.

This book grew out of a series of lecture notes for a course on Ramsey theory taught in the MASS program of the Pennsylvania State University. It was an intense and rewarding experience, and the authors hope this book conveys some of the spirit of that semester back in the fall of 2011.

It seems appropriate to close this introduction with a few words on the namesake of Ramsey theory. Frank Plumpton Ramsey (1903–1930) was a British mathematician, economist, and philosopher. A prodigy in many fields, Ramsey went to study at Trinity College Cambridge when he was 17 as a student of economist John Maynard Keynes. There, philosopher Ludwig Wittgenstein also served as a mentor. Ramsey was largely responsible for Wittgenstein's *Tractatus Logico-Philosophicus* being translated into English, and the two became friends.

Ramsey was drawn to mathematical logic. In 1928, at the age of 25, Ramsey wrote a paper regarding consistency and decidability. His paper, *On a problem in formal logic*, primarily focused on solving certain problems of axiomatic systems, but in it can be found a theorem that would become one of the crown jewels of combinatorics.

Given any r, n, and μ we can find an m_0 such that, if $m \geq m_0$ and the r-combinations of any Γ_m are divided in any manner into μ mutually exclusive classes C_i ($i = 1, 2, \ldots, \mu$), then Γ_m must contain a sub-class Δ_n such that all the r-combinations of members of Δ_n belong to the same C_i. [**53**, Theorem B, p. 267]

Ramsey died young, at the age of 26, of complications from surgery and sadly did not get to see the impact and legacy of his work.

Acknowledgment. The authors would like to thank Jennifer Chubb for help with the manuscript and for many suggestions on how to improve the book.

State College, Pennsylvania Matthew Katz
April 2018 Jan Reimann

Chapter 1

Graph Ramsey theory

1.1. The basic setting

Questions in Ramsey theory come in a specific form: For a desired property, how large must a finite set be to ensure that if we break up the set into parts, at least one part exhibits the property?

Definition 1.1. Given a non-empty set S, a *finite partition* of S is a collection of subsets $S_1, \ldots, S_r$ such that the union of the subsets is S and their pairwise intersections are empty, i.e. each element of S is in exactly one subset S_i.

A set partition is the mathematical way to describe splitting a larger set up into multiple smaller sets. In studying Ramsey theory, we often think of a partition as a *coloring* of the elements where we distinguish one subset from another by painting all the elements in each of the subsets S_i the same color, each S_i having a distinct color.

Of course, the terms "paint" and "color" should be taken abstractly. If we are using two colors, it doesn't matter if we call our colors "red" and "blue" or "1" and "2". To express things mathematically, any finite set of r colors can be identified with the set of integers $[r] := \{1, 2, \ldots, r\}$. Therefore, a partition of a set S into r subsets can be represented by a function c, where

$$c : S \to [r].$$

We will call these functions r-colorings. If we have a subset $S' \subset S$ whose elements all have the same color, we call that subset *monochromatic*. Equivalently, we can say that a subset S' is monochromatic if it is contained entirely within one subset of the set partition, or if the coloring function c restricted to S' is constant.

It is using colorings that "Ramsey-type" questions are usually phrased: How many elements does a set S need so that given any r-coloring of S (or of collections of subsets of S), we can find a monochromatic subset of a certain size and with a desired property?

A fundamental example appears in Ramsey's 1928 paper [53]:

Is there a large enough number of elements a set S needs to have to guarantee that given any r-coloring on $[S]^p$, the set of p-element subsets of S, there will exist a monochromatic subset of size k?

Ramsey showed that the answer is "yes". We will study this question throughout this chapter and prove it in full generality in Section 1.6.

Essential notation. Questions and statements in Ramsey theory can get somewhat complicated, since they will often involve several parameters and quantifiers. To help remedy this, a good system of notation is indispensable. As already mentioned, $[S]^p$ denotes the set of subsets of S of size p, where $p \geq 1$, that is,

$$[S]^p = \{T : T \subseteq S, |T| = p\}.$$

S will often be of the form $[n] = \{1, \ldots, n\}$, and to increase readability, we write $[n]^p$ for $[[n]]^p$. Note that if $|S| = n$, then $|[S]^p| = \binom{n}{p}$.

The *arrow notation* was introduced by Erdős and Rado [15]. We write

$$N \longrightarrow (k)^p_r$$

to mean that

if $|S| = N$, then every r-coloring of $[S]^p$ has a monochromatic subset of size k.

We will be dealing with colorings of sets of all kinds. For example, $c : [\mathbb{N}]^3 \to \{1, 2, 3, 4\}$ means that we have a 4-coloring of the set of

1.1. The basic setting

three-element subsets of $\mathbb{N}$. Formally, we would have to write such functions as $c(\{a_1, a_2, a_3\})$, but to improve readability, we will use the notation $c(a_1, a_2, a_3)$ instead.

The pigeonhole principle. The most basic fact about partitions of sets, as well as a key combinatorial tool, is the pigeonhole principle, often worded in terms of objects and boxes.

If n objects are put into r boxes where $n > r$, then at least one box will contain at least 2 objects.

In arrow notation,

$$n \longrightarrow (2)_r^1 \text{ whenever } n > r.$$

The pigeonhole principle seems obvious; if the r boxes have at most one object, then there can be at most r objects. However, in its simplicity lies a powerful counting argument which will form the backbone of many of the arguments in this book. It is believed that the first time the pigeonhole principle was explicitly formulated was in 1834 by Dirichlet.[1]

We can rephrase the pigeonhole principle in terms of set partitions: If a set with n elements is partitioned into r subsets where $n > r$, then at least one subset will contain at least 2 elements. From our point of view, the pigeonhole principle can be seen as the first Ramsey-type theorem: It asserts the existence of a subset with more than one element, provided n is "large enough".

The pigeonhole principle can be strengthened in the following way:

Theorem 1.2 (Strong pigeonhole principle). *If a set with n elements is partitioned into r subsets, then at least one subset will contain at least $\lceil \frac{n}{r} \rceil$ elements.*

As usual, $\lceil \frac{n}{r} \rceil$ is the least integer greater than or equal to $\frac{n}{r}$. Again, the proof is clear; if all r subsets have less than $\lceil \frac{n}{r} \rceil$ elements, then there would be fewer than n elements in the set.

---

[1] The pigeonholes in the name of the principle refer to a drawer or shelf of small holes used for sorting mail, and are only metaphorically related to the homes of rock doves. It is interesting to note that Dirichlet might have had these sorts of pigeonholes in mind as his father was the postmaster of his city [**5**].

The strong pigeonhole principle completely answers the Ramsey-type question, "how large does a set S need to be so that any r-coloring of S has a monochromatic subset of size at least k?" The answer is that N must be at least $r(k-1)+1$, and any smaller would be too few elements. We can write this result in arrow notation.

Theorem 1.3 (Strong pigeonhole principle, arrow notation).

$$N \longrightarrow (k)_r^1 \quad \text{if and only if} \quad N \geq r(k-1)+1.$$

In this case, we are able to get an exact cut-off of how large the set needs to be; however, we will see that getting exact answers to Ramsey-type questions will not always be easy, or even possible.

While the pigeonhole principle is a rather obvious statement in the finite realm, its infinite versions are not trivial and require the development of a theory of infinite sizes (*cardinalities*). We will do this in Chapter 2.

Exercise 1.4. Prove that any subset of size $n+1$ from $[2n]$ must contain two elements whose sum is $2n+1$.

1.2. The basics of graph theory

We want to move from coloring single elements of sets to coloring two-element subsets, that is, colorings on $[S]^2$. This is when the true nature of Ramsey theory starts to emerge.

Thanks to Euler [**16**], we have a useful geometric representation for subsets of $[S]^2$: combinatorial graphs. Given a subset of $[S]^2$, for each element of S you can draw a dot, or *vertex*, and then connect two dots by a line segment, or *edge*, if the pair of corresponding elements is in your subset. This sort of configuration is called a *combinatorial graph*.

For those unfamiliar with graph theory, this section will present the basic ideas from graph theory that will be needed in this book. For more background on graph theory, there are a number of excellent textbooks, such as [**4, 12**].

1.2. The basics of graph theory

Definition 1.5. A (**combinatorial**[2]) **graph** is an ordered pair $G = (V, E)$ where V, the **vertex set**, is any non-empty set and E, the **edge set**[3], is any subset of $[V]^2$.

The size of the vertex set is called the **order** of the graph G and is denoted by $|G|$. A graph may be called **finite** or **infinite** depending on the size of its vertex set. In this chapter we will deal exclusively with finite graphs, those with finite vertex sets. In the next chapter, we will encounter infinite graphs. Figure 1.1 shows an example of a finite graph with $V = \{1, 2, 3, 4, 5\}$.

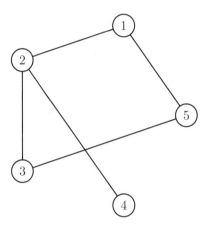

Figure 1.1. A graph with $V = \{1, 2, 3, 4, 5\}$ and $E = \{\{1,2\}, \{1,5\}, \{2,3\}, \{2,4\}, \{3,5\}\}$

The actual elements of the vertex set are often less important than its cardinality. Whether the vertex set is $\{1, 2, 3, 4, 5\}$ or $\{a, b, c, d, e\}$ carries no importance for us, as long as the corresponding graph is essentially the same. Mathematically, "essentially the same" means that the two objects are isomorphic. Two graphs $G = (V, E)$ and $G' = (V', E')$ are **isomorphic**, written $G \cong G'$, if there is a bijection

---

[2] The adjective *combinatorial* is used to distinguish this type of graph from the graph of a function. It is usually clear from the context which type of graph is meant, and so we will just speak of "graphs".

[3] Note that the definition of the edge set is not standard across all texts. Other authors may call the graphs we use **simple** graphs to emphasize that our edge set does not allow multiple edges between vertices or edges which begin and end at the same vertex, while theirs do.

φ between V and V' such that $\{v,w\}$ is an edge in E if and only if $\{\varphi(v), \varphi(w)\}$ is an edge in E'. In Figure 1.2, we see two isomorphic graphs. Although they might look rather different at first glance, mapping $1 \mapsto C$, $2 \mapsto B$, $3 \mapsto D$, and $4 \mapsto A$ transforms the left graph into the right one.

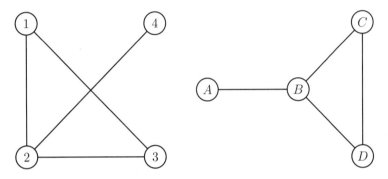

Figure 1.2. Two isomorphic graphs

We say that two vertices u and v in V are **adjacent** if $\{u,v\}$ is in the edge set. In this case, u and v are the **endpoints** of the edge. Since our edges are unordered pairs[4], adjacency is a symmetric relationship, that is, if u is adjacent to v, then v is also adjacent to u.

Given a vertex v of a graph, we define the **degree** of the vertex as the number of edges connected to the vertex and denote it by $\deg(v)$. In a finite graph, the number of edges will also be finite, and so must the degree of every vertex. However, in an infinite graph, it is possible that $\deg(v) = \infty$. In either case, each edge contributes to the degree of exactly two vertices, and so we get the *degree-sum formula*:

$$\sum_{v \in V} \deg(v) = 2|E|.$$

We say that two graphs with the same vertex set, $G_1 = (V, E_1)$ and $G_2 = (V, E_2)$, are **complements** if the edge sets E_1 and E_2 are complements (as sets) in $[V]^2$. This means that if G_1 and G_2 are

---

[4] Another possible definition would be that the edge set is a set of *ordered* pairs. The result would be a graph where each edge has a "direction" associated with it, like a one-way street. Graphs whose edges are ordered pairs are called **directed graphs**.

1.2. The basics of graph theory 7

complements, then u and v are adjacent in G_2 if and only if they are *not* adjacent in G_1, and vice versa.

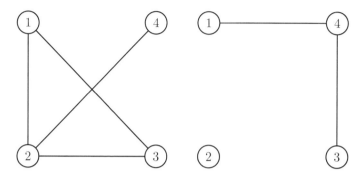

Figure 1.3. The two graphs are complements of each other.

Subgraphs, paths, and connectedness. Intuitively, the concepts of subgraph and path are easy to describe. If you draw a graph and then erase some of the vertices and edges, you get a *subgraph*. If you start at one vertex in a graph and then "walk" along the edges, tracing out your motion as you go from vertex to vertex, you have a *path*. In both cases, we are talking about restricting the vertex and/or edge sets of our graphs.

Given a graph $G = (V, E)$, if we choose a subset V' of V, we get a corresponding subset of E consisting of all the edges with both endpoints in V'. We call this subset the *restriction* of E to V' and denote it by $E|_{V'}$; formally, $E|_{V'} := E \cap [V']^2$.

Definition 1.6.

(i) Given two graphs $G = (V, E)$ and $G' = (V', E')$, if $V' \subseteq V$ and $E' \subseteq E|_{V'}$, then G' is a **subgraph** of G.

(ii) Given a graph $G = (V, E)$, if $V' \subseteq V$, then $G|_{V'} := (V', E|_{V'})$ is the **subgraph induced by** V'.

Geometrically, an induced subgraph results in choosing a set of vertices in the graph and then erasing all the other vertices and any edge whose endpoint is a vertex you just erased.

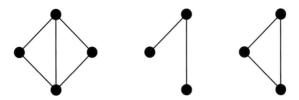

Figure 1.4. A graph $G = (V, E)$ is shown on the left. The middle graph is a subgraph of G, but not an induced subgraph, while the graph on the right is an induced subgraph of G.

Definition 1.7. A path $P = (V, E)$ is any graph (or subgraph) where $V = \{x_0, x_1, \ldots, x_n\}$ and $E = \{\{x_0, x_1\}, \{x_1, x_2\}, \ldots, \{x_{n-1}, x_n\}\}$.

Note that the definition of path requires all vertices x_i along the path to be distinct—along a path, we can visit each vertex only once. The size of the edge set in a path is called the **length** of the path. We allow paths of length 0, which are just single vertices. Rather than as a graph, we can also think of a path as a finite sequence of vertices which begin at x_0 and end at x_n. If $n \geq 2$ and x_0 and x_n are adjacent, we can extend the path to a **cycle** or **closed path**, beginning and ending at x_0.

If there exists a path that begins at vertex u and ends at vertex v, then we say u and v are **connected**. Connectedness is a good example of an equivalence relation:

- it is *reflexive*—every vertex u is connected to itself (by a path of length 0);
- it is *symmetric*—if u is connected to v, then v is connected to u (we just reverse the path);
- it is *transitive*—if u is connected to v and v is connected to w, then u is connected to w (intuitively by concatenating the two paths, but a formal proof would have to be more careful, since the paths could share edges and vertices, so we would not be able to concatenate them directly).

Recall the general definition of an equivalence relation. A binary relation R on a set X is an **equivalence relation** if for all $x, y, z \in X$,

(E1) $x R x$,

1.2. The basics of graph theory

(E2) xRy implies yRx, and

(E3) if xRy and yRz then xRz.

Every equivalence relation partitions its underlying set into **equivalence classes**;

$$[x]_R = \{yRx : y \in X\}$$

denotes the equivalence class of x. The connectedness relation partitions the vertex set into equivalence classes called **connected components** of the graph. We call a graph **connected** if it has only one connected component, that is, if any vertex is accessible to any other vertex via a path.

Exercise 1.8. Prove that a graph of order n which has more than

$$\frac{(n-1)(n-2)}{2}$$

edges must be connected.

Complete and empty graphs. Given an integer $n \geq 1$, we define the **complete graph of order** n, K_n, to be the unique graph (up to isomorphism) on n vertices where every pair of vertices has an edge between them; that is, $K_n \cong ([n], [n]^2)$. Every graph on N vertices can be viewed as a subgraph of K_N.

Figure 1.5. The complete graphs K_4, K_5, and K_6

The number of edges in K_n is $\binom{n}{2} = \frac{n(n-1)}{2}$. Although this is obvious from the definition of the binomial coefficient, it can also be shown using the vertex-sum formula: Since all of the vertices have to be adjacent, the degree of each of the n vertices must be $n-1$, and so

$$\sum_{v \in V} \deg(v) = n(n-1) = 2|E|.$$

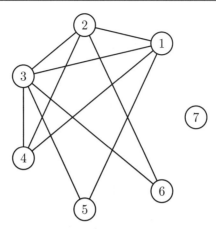

Figure 1.6. The set {1,2,3,4} forms a 4-clique, whereas the set {5,6,7} is independent.

The complete graph is on one extreme end of a spectrum where every possible edge is included in the edge set. On the other end, we would have the graph where none of the vertices are adjacent. The edge set of this graph is empty, and so we call it the **empty graph**. Note that the complete and empty graphs on n vertices are complements of each other.

Given a graph $G = (V, E)$, if V' is a subset of V and $G|_{V'}$ is complete, then we say that V' is a **clique**. Specifically, if V' has order k, then V' is a k-**clique**.

On the other hand, if $G|_{V'}$ is an empty graph, then we say that V' is **independent** (see Figure 1.6).

Bipartite and k-partite graphs. Let $G = (V, E)$ be a graph, and let V be partitioned into V_1 and V_2; that is, $V_1 \cup V_2 = V$ and $V_1 \cap V_2 = \emptyset$. Consider the case where $E \subseteq V_1 \times V_2 \subset [V]^2$, so that each edge has one endpoint in V_1 and one endpoint in V_2. Such a graph is called a **bipartite graph**. An equivalent definition is that the vertex set can be partitioned into two independent subsets.

Notice that in the right example in Figure 1.7, no more edges could have been added to that graph without destroying its bipartiteness; every vertex in the left column is adjacent to every vertex

1.2. The basics of graph theory

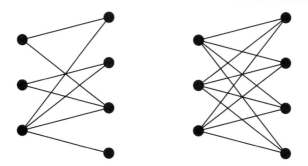

Figure 1.7. Two bipartite graphs

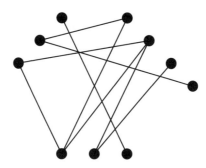

Figure 1.8. A 3-partite graph

in the right column. If $G = (V_1 \cup V_2, E)$ is a bipartite graph where $|V_1| = n$ and $|V_2| = m$, and every vertex in V_1 is adjacent to every vertex in V_2, then G is the **complete bipartite graph of order** n, m, and we denote it by $K_{n,m}$.

Exercise 1.9. Describe the complement of $K_{n,m}$.

We can generalize the definition of bipartite to say that a graph is k-**partite** if the vertex set can be partitioned into k independent subsets.

If G is a k-partite graph whose vertex set is partitioned into V_1 through V_k, where $|V_i| = n_i$ and each vertex of V_i is adjacent to every vertex in all the V_j with $j \neq i$, then our graph is the **complete k-partite graph of order** $n_1, \ldots, n_k$, and we denote it by $K_{n_1, \ldots, n_k}$.

Exercise 1.10. Prove that the total number of edges in $K_{n_1,\ldots,n_k}$ is $\sum_{1 \le i < j \le k} n_i n_j$.

Trees. A **tree** is a connected graph which contains no cycles. Trees show up in many fields of math, but also across many disciplines, from decision trees to phylogenetic trees.

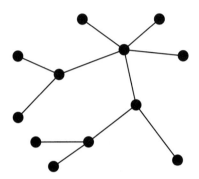

Figure 1.9. A tree

Theorem 1.11. *A graph is a tree if and only if there is a unique path between any two vertices.*

Proof. Assume there are two paths that connect vertices u and v. We may also assume that the two paths do not share any vertices except for u and v, since in that case we could replace u or v by the first vertex that the two paths share (and obtain two shorter paths to which we could apply the argument). We can create a cycle by concatenating the paths in the following way: If the first path goes through the vertices $u, x_1, \ldots, x_n, v$ and the second path goes through the vertices $u, y_1, \ldots, y_m, v$, take $u, x_1, \ldots, x_n, v, y_m, \ldots, y_1, u$. Therefore, a tree will always have a unique path between any two vertices.

On the other hand, assume that we have a graph G which is not a tree. This means that either G is disconnected or there is a cycle in G. If G is disconnected, then there are two vertices u and v that are in different connected components and so do not have a path between them. If G has a cycle, the cycle contains at least two vertices, u' and v'. This path can then be decomposed into two paths, one from u'

1.2. The basics of graph theory 13

to v' and one from v' to u', which means that there is more than one path between the two vertices. Therefore, any graph in which there is a unique path between any two vertices would be a connected graph with no cycles, a tree. □

Exercise 1.12. Prove that a graph is a tree if and only if removing an edge makes the graph disconnected.

Exercise 1.13. Prove that a connected graph on n vertices is a tree if and only if it has $n-1$ edges.

The fact that two vertices are connected by a unique path lets us organize a tree in a hierarchical manner. We designate one vertex in a tree to be the **root** of the tree. Then, any other vertex in the tree with degree 1 is called a **leaf**. Once a root has been chosen, we can reorient any tree with the root at the bottom and all the leaves at the top, like a real tree.

After choosing a root vertex, we can **partially order** the vertices of a tree, based on their distance from the root. Given a vertex v, consider the unique path from the root r to v. If this path goes through a vertex u, then we say that v is a **successor** of u, or that u is a **predecessor** of v, and we write $u < v$.

This order is *partial* in the sense that if $u \neq v$ are not on the same path from the root, they are not comparable, that is, neither $u < v$ nor $v < u$.

The root is the unique vertex which is a predecessor of all other vertices. A path from the root r to v can be represented as the sequence of vertices

$$r = v_0 < v_1 < \cdots < v_{n-1} < v_n = v,$$

where each v_i is an immediate successor of v_{i-1}.

The induced partial order is an important aspect of trees, and we will return to it in Section 2.2.

Graph colorings. Graph colorings generally come in two varieties: edge colorings and vertex colorings. Since we are using graphs as a means of illustrating subsets of $[V]^2$ as edges, we will be primarily interested in edge colorings. Given a graph $G = (V, E)$, an r-**edge coloring**, or simply r-coloring, of G is a function $c : E \to [r]$.

Any graph $G = (V, E)$ on N vertices *induces* a 2-coloring on K_N in the following way: If two vertices are adjacent in G, paint their edge in K_N blue; otherwise paint it red.

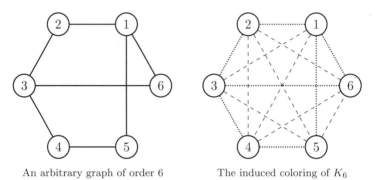

An arbitrary graph of order 6   The induced coloring of K_6

Figure 1.10. Translating between arbitrary graphs and edge colorings of a complete graph ($\cdots$ = blue, $--$ = red)

The induced coloring can also be seen as a set partition of $[V]^2$ into two complementary parts E_1 and E_2; we can color the edges in E_1 blue and those in E_2 red, and (V, E_1) and (V, E_2) will be graph complements in K_N. Likewise, we can view an r-coloring as representing a set partition of $[V]^2$ into r parts, resulting in r mutually exclusive graphs.

1.3. Ramsey's theorem for graphs

Suppose you arrive at a party. As you browse the room, you see some familiar faces, whereas others are complete strangers to you. As you make your way to the buffet, a curious thought enters your mind: *Will there be at least three people who all know each other? And if not, are there three people who have never met before?*

As you are mathematically inclined, you notice your question has a graph-theoretic formulation: We can represent each guest by a vertex. If two guests know each other, we draw a blue line between them, and if they have never met, we draw a red line between them.

1.3. Ramsey's theorem for graphs

What we have is a representation of the party as a 2-coloring of K_n, where n is the number of guests.

Now the original question becomes:

If we 2-color the edges of K_n, can we find a red or a blue triangle?

Exercise 1.14. Show that if only five people attend, the answer to your question can be negative. In other words, find a 2-coloring of K_5 without a monochromatic triangle.

So let us assume there are at least six people attending and consider any 2-coloring of K_6. Let us call the vertices $v_1, v_2, \ldots, v_6$ and consider, without loss of generality, the first vertex v_1. Vertex v_1 is connected to five other vertices. If we let R be the set of vertices connected to v_1 by a red edge and let B be the set of vertices connected to v_1 by a blue edge, then by the pigeonhole principle, either $|R| \geq 3$ or $|B| \geq 3$; we will assume $|R| \geq 3$. If any two elements in R, say v_2 and v_3, are connected by a red edge, then $v_1, v_2,$ and v_3 are the vertices of a red triangle. On the other hand, if all the elements in R are connected by blue edges, then we have a blue triangle since there are at least three vertices in R. In either case, we have a monochromatic triangle (Figure 1.11).

In arrow notation, we just showed that $6 \longrightarrow (3)_2^2$. (Keep in mind that the subscript is denoting that we are using 2 colors, and the superscript is denoting that we are coloring 2-element subsets.)

Surely this result will hold if our original complete graph was on more than six vertices; simply pick six of the vertices and consider the induced subgraph on those vertices, which is necessarily K_6, and then use the result. Since we previously showed that five vertices is not enough, we have proven the following.

Proposition 1.15. $N \longrightarrow (3)_2^2$ *if and only if* $N \geq 6$.

We should note how important the pigeonhole principle was to our argument, specifically that $5 \longrightarrow (3)_2^1$. Note also that while we end up with a monochromatic subset, we do not know in advance which color it will have. This only becomes clear during the process of finding it.

1. Graph Ramsey theory

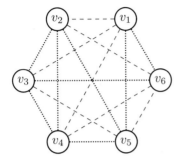

We start with a 2-colored K_6. Pick an arbitrary vertex, for example v_1.

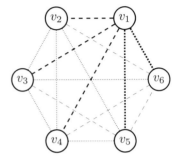

Three edges connecting v_1 to the other vertices are "red" (dashed). The vertices at the other end of these edges are $v_2, v_3,$ and v_4.

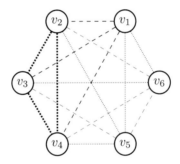

The edges between $v_2, v_3,$ and v_4 are all "blue" (dotted), yielding the desired monochromatic K_3. Were one of the edges red, its vertices, together with v_1, would give rise to a red (dashed) triangle.

Figure 1.11. Proving $6 \to (3)_2^2$

If the party has more guests, can we find even larger cliques (of mutual friends or mutual strangers)? This is the subject of Ramsey's theorem for graphs.

Theorem 1.16 (Ramsey's theorem for 2-colored graphs)**.** *For any $k \geq 2$, there exists some integer N such that any 2-coloring of a graph*

1.3. Ramsey's theorem for graphs

of at least N vertices contains a complete monochromatic subgraph on k vertices.

Proposition 1.15 was a special case of this statement. There, we saw that for $k = 3$ we can choose $N = 6$. The dual form of the theorem, using cliques and independent sets, reads as follows.

Corollary 1.17 (Ramsey's theorem for graphs, dual form). *For any $k \geq 2$, there exists some integer N such that any graph of at least N vertices contains a complete subgraph on k vertices or an independent subgraph on k vertices.*

In the course of this book, we will encounter several proofs of Theorem 1.16. We start with the one that is not the most elegant but arguably the most elementary, as it uses the pigeonhole principle in an almost "brute force" way.

Proof of Theorem 1.16. Consider K_N, the complete graph on N vertices, where we think of N for now as a sufficiently large integer. Suppose the edges of K_N are 2-colored, red and blue.

We will construct a monochromatic K_k in two stages. In the first stage, we pick a sequence of vertices

$$v_1, v_2, v_3, \ldots$$

such that v_i is connected to all following vertices by an edge of the same color. This color, however, may change from vertex to vertex. In the second stage, we select a subsequence from the v_i that yields a monochromatic K_k.

Stage 1: We start by picking an arbitrary vertex v_1 and consider the $N-1$ remaining vertices. These are split into two subsets: the ones we call the *blue* vertices because they connect to v_1 via a blue edge, and the *red* vertices that connect via a red edge. By the pigeonhole principle, there are either $\lceil (N-1)/2 \rceil$ red or $\lceil (N-1)/2 \rceil$ blue vertices. Call the color for which this holds c_1 and let V_2 be the set of color-c_1 vertices, that is, vertices that connect to v_1 via an edge of color c_1. V_2 induces a subgraph $G_2 = (V_2, E_2)$ (E_2 contains all edges between vertices in V_2).

We now continue working in G_2 and repeat the whole process by choosing an arbitrary vertex $v_2 \in V_2$. G_2 has at least $\lceil (N-1)/2 \rceil$ vertices, so again by the pigeonhole principle, there are at least

$$\left\lceil \frac{\lceil \frac{N-1}{2} \rceil - 1}{2} \right\rceil \geq \left\lceil \frac{N-3}{4} \right\rceil$$

vertices in G_2 that are connected to v_2 by an edge of the same color. Call this color c_2. We collect the vertices adjacent to v_2 via these edges in the set V_3, which in turn induces a new subgraph G_3 of G_2.

If we continue this process, we get a sequence of vertices

$$v_1, v_2, v_3, \ldots, v_t$$

and a sequence of colors

$$c_1, c_2, c_3, \ldots, c_t$$

(and a sequence subgraphs $G = G_1 \supset G_2 \supset G_3 \supset \cdots \supset G_t$). Since we "take out" at least one vertex in each step, the process will terminate after finitely many, say t, steps, because the graph we started with has only finitely many vertices. By the way we chose these sequences, they have the following property:

For any vertex v_i in the sequence, all later vertices v_j with $j > i$ are connected to v_i by an edge of color c_i.

We are not quite done yet, because the colors c_i can be different for each vertex. If these colors were all identical, the whole sequence $v_1, v_2, \ldots$ would induce a complete monochromatic subgraph.

Stage 2: We use the pigeonhole principle one more time. Among the color sequence $c_1, c_2, c_3, \ldots, c_t$, one color must occur at least half the time. Let c be such a color. Now consider all the vertices v_i for which $c_i = c$. Collect them in a vertex set V_c. We claim that V_c induces a complete monochromatic subgraph of color c. Let v and w be two vertices in V_c. They have entered the sequence at different stages, say $v = v_{i_0}$ and $w = v_{i_1}$, where $i_0 > i_1$. But, by definition of V_c, every edge between v_{i_0} and a later vertex is of color $c_{i_0} = c$. Since v and w were arbitrary vertices of V_c, all edges between vertices in V_c have color c.

1.3. Ramsey's theorem for graphs

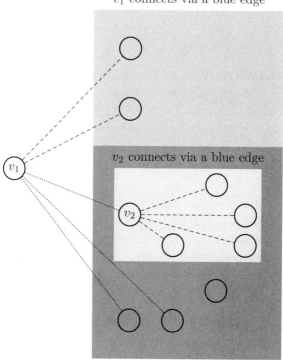

Figure 1.12. Selecting the sequence of nodes $v_1, v_2, \ldots$

We have hence a set of vertices V_c of cardinality $\geq \lceil t/2 \rceil$ that induces a monochromatic subgraph. We want a complete graph with k vertices. That means we need to choose N large enough so that $\lceil t/2 \rceil \geq k$.

A straightforward induction yields that for $s < t$,

$$|G_{s+1}| \geq \left\lceil \frac{N - (2^s - 1)}{2^s} \right\rceil \geq \frac{N}{2^s} - 1.$$

This means that Stage 1 has at least $\lfloor \log_2 N \rfloor$ have steps, i.e. $t \geq \lfloor \log_2 N \rfloor$. This in turns implies that if we let

$$N = 2^{2k},$$

we get a monochromatic complete subgraph of size
$$\frac{\log_2 2^{2k}}{2} = k.$$

$\square$

Exercise 1.18. Use Ramsey's theorem for graphs to show that for every positive integer k there exists a number $N(k)$ such that if $a_1, a_2, \ldots, a_{N(k)}$ is a sequence of $N(k)$ integers, it has a non-increasing subsequence of length k or a non-decreasing subsequence of length k. Show further that $N(k+1) > k^2$.

(*Hint:* Find a way to translate information about sequences rising or falling into a graph coloring.)

Multiple colors. In our treatment so far we have dealt with 2-colorings of complete graphs, mainly because of the nice correspondence between monochromatic complete subgraphs and cliques/independent sets as described in Figure 1.10. But Theorem 1.16 can be extended to hold for any finite number of colors.

Theorem 1.19. *For any $k \geq 2$ and for any $r \geq 2$ there exists some integer N such that any r-coloring of a complete graph of at least N vertices contains a complete monochromatic subgraph on k vertices.*

The proof of this theorem is very similar to the proof of Theorem 1.16. One can in fact go through it line by line and adapt it to the case of r colors; we will simply give the broad strokes.

In Stage 1, we choose a vertex v_0 and partition the other vertices by the color of the edge they share with v_0. The largest of the subsets will have size at least $1/r$ times the first set. If we started with N vertices, this process will terminate in t steps where $t \geq \lfloor \log_r N \rfloor$. So if we choose $N = r^{rk}$, then our process will terminate in rk steps or more.

Now, in Stage 2, we have a sequence of rk vertices, each connected to all of the following vertices by the same color. This color, however, can differ from vertex to vertex, one of r many colors. Another application of the pigeonhole principle as in the proof of Theorem 1.16 gives us a set of k vertices that *all* connect to all subsequent vertices by a *single* color. These vertices form a monochromatic k-clique. Therefore, $r^{rk} \longrightarrow (k)_r^2$.

1.4. Ramsey numbers and probabilistic method

Exercise 1.20. Prove *Schur's theorem*: For every positive integer k there exists a number $M(k)$ such that if the set $\{1, 2, \ldots, M(k)\}$ is partitioned into k subsets, at least one of them contains a set of the form $\{x, y, x + y\}$.

(*Hint:* Consider a complete graph on vertices $\{0, 1, 2, \ldots, M\}$, where M is an integer. Devise a k-coloring of the graph such that the color of an edge reflects an arithmetic relation between the vertices with respect to the k-many partition sets.)

1.4. Ramsey numbers and the probabilistic method

As we described in the introduction, Ramsey's theorem is remarkable in the sense that it *guarantees regular substructures in any sufficiently large graph*, no matter how "randomly" that graph is chosen. But how large is "sufficiently large"?

Definition 1.21. The **Ramsey number** $R(k)$ is the least natural number N such that $N \longrightarrow (k)_2^2$.

From our proof of Ramsey's theorem (Theorem 1.16) we obtain an upper bound $R(k) \leq 2^{2k}$. Is this bound sharp? It gives us $R(3) \leq 64$, while we already saw in the previous section that $R(3) = 6$ (Proposition 1.15). The proof yields $R(10) \leq 2^{20} = 1\,048\,576$, but it was shown in 2003 [**59**] that $R(10) \leq 23\,556$; in fact the true value of $R(10)$ could still be much lower. It seems that we have just been a little "wasteful" when proving Ramsey's theorem for graphs, using way more vertices in our argument than actually required. This leads us to the question of finding better *upper and lower bounds for Ramsey numbers*, a notoriously difficult problem.

In 1955, Greenwood and Gleason published a paper [**27**] which provided the first exact Ramsey numbers for $k > 3$, as well as answers to generalizations of the problem. Their proof is a nice example of the fact that sometimes it is easier to prove a more general statement.

Instead of finding monochromatic cliques of the same order, one can break the symmetry and look for cliques of different orders—for example, either a red m-clique or a blue n-clique. We can extend our arrow notation by saying that

$$N \longrightarrow (m,n)_2^2$$

if every 2-coloring of a complete graph of order N has either a red complete subgraph on m vertices or a blue complete subgraph on n vertices.

Definition 1.22. The **generalized Ramsey number** $R(m,n)$ is the least integer N such that $N \longrightarrow (n,m)_2^2$.

The Ramsey number $R(k)$ introduced at the beginning of this section is equal to the *diagonal* generalized Ramsey number $R(k,k)$.

It is clear that $R(m,n)$ exists for every pair m,n, since $R(m,n) \leq R(k)$ when $k = \max(m,n)$. On the other hand, $R(m,n) \geq R(k)$ when $k = \min(m,n)$. Since one can swap the colors on all the edges from red to blue or from blue to red, $R(m,n) = R(n,m)$.

Some values of $R(m,n)$ are known. We put $R(m,1) = R(1,n) = 1$ for all m,n, since K_1 has no edges. Any single vertex is, trivially, a monochromatic 1-clique of any color.[5] If we consider finding $R(m,2)$, we are looking for either a red m-clique or a blue 2-clique. But if any edge in a coloring is blue, its endpoint vertices form a monochromatic 2-clique. If a graph does not have a blue edge, it must be completely red. Therefore $R(m,2) = m$, and similarly $R(2,n) = n$.

Greenwood and Gleason proved a recursive relation between the Ramsey numbers $R(m,n)$. As their result does not assume the existence of *any* diagonal Ramsey number $R(k)$, we in particular obtain a new, elegant proof of Theorem 1.16.

Theorem 1.23 (Greenwood and Gleason). *The Ramsey numbers $R(m,n)$ exist for all $m,n \geq 1$ and satisfy*

(1.1) $$R(m,n) \leq R(m-1,n) + R(m,n-1)$$

for all $m,n \geq 2$.

Proof. We will proceed by *simultaneous induction* on m and n, meaning that we deduce that $R(m,n)$ exists using the fact that both $R(m,n-1)$ and $R(m-1,n)$ exist. To be more specific, we can arrange

---

[5] We use this definition only to aid us in induction in the following proofs. We will generally ignore this case in discussion.

1.4. Ramsey numbers and probabilistic method

the pairs m, n in a matrix and then use the truth of the statement for values in the ith diagonal of the matrix to prove the statement in the $(i + 1)$st diagonal. In each diagonal, the sum $m + n$ is constant, and so we can view the simultaneous induction on m and n as a standard induction on the value $m + n$.

For the base case of the induction ($m = n = 2$), (1.1) follows easily from the fact that $R(2, n) = n$ and $R(m, 2) = m$. For the inductive step, let $N = R(m - 1, n) + R(m, n - 1)$. Consider a 2-colored K_N and let v be an arbitrary vertex. Define V_{red} and V_{blue} to be the vertices connected to v via a red edge or a blue edge, respectively. Then

(1.2) $\quad |V_{\text{red}}| + |V_{\text{blue}}| = N - 1 = R(m - 1, n) + R(m, n - 1) - 1.$

By the pigeonhole principle, either $|V_{\text{red}}| \geq R(m - 1, n)$ or $|V_{\text{blue}}| \geq R(m, n - 1)$. For if $|V_{\text{red}}| \leq R(m - 1, n) - 1$ and $|V_{\text{blue}}| \leq R(m, n - 1) - 1$, then

$$|V_{\text{red}}| + |V_{\text{blue}}| \leq R(m - 1, n) + R(m, n - 1) - 2.$$

We want to argue that in either case, K_N has either a complete red subgraph on m vertices or a complete blue subgraph on n vertices. Let us first assume $|V_{\text{red}}| \geq R(m - 1, n)$. From the definition of the Ramsey number, V_{red} has either a complete red subgraph on $m - 1$ vertices or a complete blue subgraph on n vertices. In the second case we are done right away. In the first case, we can add our chosen v to the set V_{red} and the new red subgraph is complete on m vertices. The argument for $|V_{\text{blue}}| \geq R(m, n - 1)$ is similar. □

There is a certain similarity between the inequality in (1.1) and the famous relationship between binomial coefficients:

$$\binom{n}{m} = \binom{n-1}{m} + \binom{n-1}{m-1}.$$

We can exploit this recursive relation to arrive at an upper bound for $R(k)$ which is better than the one we had before.

Theorem 1.24. *For all $m, n \geq 2$, the Ramsey number $R(m, n)$ satisfies*

$$R(m, n) \leq \binom{m + n - 2}{m - 1}.$$

Proof. We have
$$R(m,2) = m = \binom{m}{m-1} = \binom{m+2-2}{m-1},$$
$$R(2,n) = n = \binom{n}{1} = \binom{2+n-2}{2-1},$$
and then, again by simultaneous induction,
$$R(m,n) \leq R(m-1,n) + R(m,n-1)$$
$$\leq \binom{m+n-3}{m-2} + \binom{m+n-3}{m-1} = \binom{m+n-2}{m-1}. \qquad \square$$

For $m = n = k$, this theorem yields the upper bound
$$R(k) \leq \binom{2k-2}{k-1}. \tag{1.3}$$
Using Stirling's approximation formula for $n!$, one can show that when n is sufficiently large, $\binom{2k-2}{k-1}$ is approximately $2^{2(k-1)}/\sqrt{\pi(k-1)}$, so the bound in (1.3) is a little better than our original bound $R(k) \leq 2^{2k}$ from the proof of Theorem 1.16.

The current (as of 2018) best known general upper bound for $R(k)$ was proved by Conlon [10] in 2009: There exists a constant C such that
$$R(k) \leq k^{-C\frac{\log(k-1)}{\log\log(k-1)}} \binom{2k-2}{k-1}.$$

If $R(m-1,n)$ and $R(m,n-1)$ are both even, we can show that the inequality of Theorem 1.23 is strict.

Proposition 1.25. *If $R(m-1,n)$ and $R(m,n-1)$ are both even, then $R(m,n) < R(m-1,n) + R(m,n-1)$.*

Proof. Assume that $R(m-1,n) = 2p$ and $R(m,n-1) = 2q$, with p and q being integers. Consider a 2-colored K_N where $N = 2p + 2q - 1$. We claim that K_N has a red K_m or a blue K_n.

Let v be any node and consider the sets $V_{\text{red}}(v)$ and $V_{\text{blue}}(v)$ for v as defined in the proof of Theorem 1.23. The following three cases are possible:

(a) $|V_{\text{red}}(v)| \geq 2p$,

(b) $|V_{\text{blue}}(v)| \geq 2q$,

1.4. Ramsey numbers and probabilistic method

(c) $|V_{\text{red}}(v)| = 2p - 1$ and $|V_{\text{blue}}(v)| = 2q - 1$.

In cases (a) and (b), we can argue as in the proof of Theorem 1.23 that a monochromatic K_m or K_n exists. Case (c) requires a counting argument, where we will use the parity assumption.

As v was arbitrary, we can carry out the above argument for *all* nodes. If for any node (a) or (b) holds, we are done. So let us assume that for every node v, (c) holds. Every edge has two ends. We identify the color of the ends with the color of the edge. Since (c) holds for every node, $(2p + 2q - 1)(2p - 1)$ of the edges have red ends. $(2p + 2q - 1)(2p - 1)$ is an odd number, but every edge has two ends, which implies an even number of red ends, leading to a contradiction. □

Some exact values. For some small values of m, n, we are able to determine $R(m, n)$ exactly. We have already mentioned that for $m, n \geq 2$, $R(m, 2) = m$ and $R(2, n) = n$. We have also proved in Section 1.3 that $R(3,3) = 6$.

Proposition 1.26. $R(3,4) = 9$ *and* $R(3,5) = 14$

Proof. From Theorem 1.23, we have that $R(3,4) \leq R(3,3) + R(2,4) = 6 + 4 = 10$, and by Proposition 1.25, the inequality is strict. So $R(3,4) \leq 9$.

Greenwood and Gleason were able to describe a graph of order 13 which has neither a red 3-clique nor a blue 5-clique, implying that $R(3,5) > 13$. Then $14 \leq R(3,5) \leq R(2,5) + R(3,4)$, and since $R(2,5) = 5$, we have $R(3,4) \geq 9$, which proves $R(3,4) = 9$; plugging that back into the inequality gives us $R(3,5) = 14$. □

Proposition 1.27. $R(4,4) = 18$.

Proof. From Theorem 1.23 and the previous proposition, we have that $R(4,4) \leq R(3,4) + R(4,3) = 18$.

To prove that 18 is the actual value of $R(4,4)$, we have to produce a 2-coloring of a K_{17} that does not have a monochromatic K_4; in fact, there is *exactly one* such coloring, which is defined using ideas from elementary number theory.

Let p be any prime which is congruent to 1 modulo 4. An integer x is a *quadratic residue modulo p* if there exists an integer z such that $z^2 \equiv x \bmod p$. We can define the **Paley graph of order** p as the graph on $[p]$ where the vertices $x, y \in \{1, 2, \ldots, p\}$ have an edge between them if and only if $x-y$ is a quadratic residue modulo p. This definition does not depend on whether we consider $x-y$ or $y-x$, since $p \equiv 1 \bmod 4$ and -1 is a square for such prime p. The set of quadratic residues is always equal in size to the set of quadratic non-residues, both sets having $\frac{p-1}{2}$ elements, as 0 is considered neither a residue nor a non-residue. The symmetry between quadratic residues and non-residues also causes the Paley graphs to be self-complementary, that is, the complement of a Paley graph is isomorphic to itself. At the root of all this lies the *law of quadratic reciprocity*, first proved by Gauss in his *Disquisitiones Arithmeticae* [18].[6]

Figure 1.13 shows the Paley graph of order 17; we can use this graph to induce a 2-coloring on K_{17}. It can be shown using a bit of elementary number theory that this graph does not have a 4-clique. Since the graph is also self-complementary, it has no independent sets of 4 vertices either. Therefore, $R(4) = 18$. □

Despite their rather regular appearance, the Paley graphs are important objects in the study of *random graphs*, as they share many statistical properties with graphs for which the edge relation is determined by a random coin toss.

Moving on to higher Ramsey numbers, can we get an exact value for $R(5)$? As both $R(3,5)$ and $R(4,4)$ are even, Proposition 1.25 gives us
$$R(4,5) \leq R(3,5) + R(4,4) - 1 = 31,$$
and hence
$$R(5,5) \leq R(4,5) + R(5,4) \leq 62.$$

However, in this case the upper bound for $R(4,5)$ turned out not to be exact. In 1995, McKay and Radziszowski [45] showed that $R(4,5) = 25$. This in turn improved the upper bound for $R(5,5)$ to $R(5,5) \leq 49$.

---

[6] For a proof of the reciprocity law as well as for further background in number theory, the book by Hardy and Wright [31] is a classic yet hardly surpassed text.

1.4. Ramsey numbers and probabilistic method

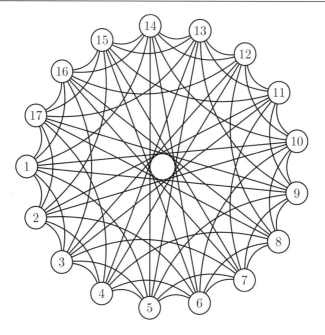

Figure 1.13. The Paley graph of order 17. The quadratic residues modulo 17 are 1, 2, 4, 8, 9, 13, 15, and 16. This graph does not contain K_4 as a subgraph.

At this point, one may ask: There are only finitely many 2-colorings of a complete graph of any finite order. Could we not cycle through all colorings of K_{48} one by one (preferably on a fast computer) and check whether each has either a red or a blue 5-clique? If not, then $R(5) = 49$. If yes, then we could test all 2-colorings of a K_{47}, and so on. Eventually we will have determined the fifth Ramsey number.

The problem with this strategy is that *there are simply too many graphs to check!* How many colorings are there? A K_{48} has $\binom{48}{2}$ = 1128 edges. Each edge can be colored in two ways, giving us

$$2^{1128} \approx 3.6 \times 10^{339}$$

colorings to check. At the time this book was written, the world's fastest supercomputer, the Cray Titan, could perform about 20×10^{15} floating-point operations per second (FLOPS). Under the unrealistic

1. Graph Ramsey theory

	$n=3$	$n=4$	$n=5$		$n=6$		$n=7$	
			low	up	low	up	low	up
$m=3$	6	9	14		18		23	
$m=4$		18	25		36	41	49	68
$m=5$			43	48	58	87	80	143
$m=6$					102	165	115	298
$m=7$							205	540

Table 1. Exact Ramsey numbers $R(m,n)$ and best known bounds for $m, n \leq 6$, from [**52**]

assumption that one coloring can be checked for monochromatic 5-cliques within one floating-point operation, it would take more than 10^{315} years to check all of them. According to current theories, Earth will be absorbed by the Sun in less than 10^{10} years.

As brute force searches seem out of our range, we will have to find more sophisticated algorithms with a significantly reduced search space and search time. In 2017, McKay and Angeltveit [**3**] indeed presented a computer verification of $R(5) \leq 48$, which is (as of early 2018) still the best upper bound for $R(5)$.

While some progress has been made, Paul Erdős's prophecy seems to have lost little of its punch:

Suppose aliens invade the earth and threaten to obliterate it in a year's time unless human beings can find the Ramsey number for red five and blue five. We could marshal the world's best minds and fastest computers, and within a year we could probably calculate the value. If the aliens demanded the Ramsey number for red six and blue six, however, we would have no choice but to launch a preemptive attack. [**25**]

Table 1 shows the current knowledge of small Ramsey numbers (as of January 2018).

1.4. Ramsey numbers and probabilistic method

More colorful Ramsey numbers. We can extend our arrow notation if we want to look for monochromatic complete subgraphs of various sizes when we have more than two colors. If we have r colors, $c_1, \ldots, c_r$, then we can write

$$N \to (n_1, \ldots, n_r)_r^2$$

if every r-coloring of a complete graph of order N has a complete subgraph with n_i vertices which is monochromatic in color c_i for some $1 \leq i \leq r$; define $R(n_1, \ldots, n_r)$ to be the associated Ramsey number.

Greenwood and Gleason showed that Theorem 1.23 generalizes to multiple colors.

Theorem 1.28. *For $n_1, \ldots, n_r \geq 1$,*

$$R(n_1, n_2, \ldots, n_r) \leq R(n_1 - 1, n_2, \ldots, n_r) + \cdots + R(n_1, n_2, \ldots, n_r - 1).$$

They also provided the first non-trivial example of an exact Ramsey number for more than two colors, by showing that $R(3, 3, 3) = 17$.

The probabilistic method. Maybe, for the case of $R(5)$, there is an object like the Paley graph for $R(4)$ that could give us a good (preferably optimal) *lower bound* on $R(5)$, or even a general construction that could give us an optimal lower bound on $R(k)$.

Erdős [14] had the idea that a *randomly colored graph* will be rather adverse to having large monochromatic subgraphs. A random graph is not a distinguished single graph, but rather a graph whose properties are determined by the principles of *probability theory*. In our case, it is the *coloring* of the edges of a complete graph that is determined randomly, say by tossing a fair coin. The *probabilistic method* determines the probability that such a *random coloring does not yield a monochromatic k-clique*. If the probability is positive, this means that such a coloring must exist. This will give us the desired lower bound, and random graphs do it rather by the fact that there is a large number of them than by a distinguished property.

Theorem 1.29 (Lower bound for $R(k)$). *For $k \geq 3$, $R(k) > 2^{k/2}$.*

Proof. The idea is to think of the coloring of the edges of a graph as a random process. If we fix a number of vertices N, we can take K_N

and color each of the $\binom{N}{2}$ edges based on a fair coin flip, say red for heads and blue for tails. Since the coin flips are independent, there will be a total of $2^{\binom{N}{2}}$ different 2-colorings of K_N, each occurring with equal probability.

Pick k vertices in your graph. There are $2^{\binom{k}{2}}$ possible 2-colorings of this subgraph and exactly 2 of them are monochromatic. Therefore, the probability of randomly getting a monochromatic subgraph on these k vertices is $2^{1-\binom{k}{2}}$. There are $\binom{N}{k}$ different k-cliques in K_N, so the probability of getting a monochromatic subgraph on any k vertices is $\binom{N}{k} 2^{1-\binom{k}{2}}$.

Now suppose $N = 2^{k/2}$. We want to show that there is a positive probability that a random coloring of K_N will have no monochromatic k-clique (and hence deduce that $R(k) > 2^{k/2}$, proving the theorem).

We bound the probability that a random coloring will give us a monochromatic k-clique from above:

$$\binom{N}{k} 2^{1-\binom{k}{2}} = \frac{N!}{k!(N-k!)} 2^{1-\binom{k}{2}}$$
$$\leq \frac{N^k}{k!} 2^{1-\binom{k}{2}} \quad \text{(since } \frac{N!}{(N-k)!} \leq N^k\text{)}$$
$$= \frac{2^{k^2/2}}{k!} 2^{1-(k^2-k)/2}$$
$$= \frac{2^{1+k/2}}{k!}.$$

If $k \geq 3$, then $\frac{2^{1+k/2}}{k!} < 1$. (This is easily verified by induction.) Therefore, it is not certain to always obtain a monochromatic clique of size k, which in turn means that there is a positive probability that a random coloring of K_N will have no monochromatic k-clique. $\square$

While Erdős was not the first to use this kind of argument, he certainly popularized it and, through his results, helped it become an important tool not only in graph theory and combinatorics but also in many other areas of mathematics (see for example [2]).

1.5. Turán's theorem

While Ramsey's theorem tells us that in a 2-coloring of a sufficiently large complete graph we always find a monochromatic clique, we do not know what color that clique will be. One would think that if there were quite a few more red edges than blue edges, we would be assured a red clique—but how many "more" red than blue do we need?

As before, rather than talking about 2-colorings of a complete graph, we can rephrase this discussion in terms of the existence of cliques. We should expect that a graph with a lot of edges should contain a complete subgraph, but what do we mean by "a lot"? This is the subject of Turán's theorem [**65**].

Theorem 1.30 (Turán's theorem). *Let* $G = (V, E)$, *where* $|V| = N$, *and let* $k \geq 2$. *If*
$$|E| > \left(1 - \frac{1}{k-1}\right)\frac{N^2}{2},$$
then G has a k-clique.

In graph theory texts, this often phrased as a result about k-clique-free graphs: If a graph has no k-clique, then it has at most $\left(1 - \frac{1}{k-1}\right)\frac{N^2}{2}$ edges. In his proof, Turán provided examples of graphs, now called *Turán graphs*, which have exactly $\left(1 - \frac{1}{k-1}\right)\frac{N^2}{2}$ edges and no k-cliques. Turán graphs are the largest graphs such that adding any edge would create a k-clique, and are therefore considered to be *extremal*. Extremal graphs, the largest (or smallest) graphs with a certain property, are the objects of interest in the field of *extremal graph theory*, for which Turán's theorem is one of the founding results.

If we begin with $k = 3$, our goal is to find a graph on N vertices with as many edges as possible but with no triangles. For this, we have to look no further than bipartite graphs. Indeed, if $\{v_1, v_2\}$ and $\{v_2, v_3\}$ are in the edge set of some bipartite graph, then v_1 and v_3 are elements of the same part, and are therefore not connected. This is true for any bipartite graph, but the complete bipartite graphs will have the most edges.

For $N = 6$, we can look at the bipartite graphs $K_{1,5}$, $K_{2,4}$, and $K_{3,3}$ and note that these graphs have 5, 8, and 9 edges respectively.

Objects in mathematics tend to achieve maxima when the sizes of parts involved are balanced. The rectangle with the largest area-to-perimeter ratio is the one where the sides have equal length. Likewise, our graph $K_{n,m}$ will have a maximal number of edges when the sizes n and m are balanced.

We can extend this example to $k > 3$ by considering complete $(k-1)$-partite graphs. It is clear by the pigeonhole principle that these graphs have no k-clique; any complete subgraph can choose only one vertex from each of the $(k-1)$ subsets of vertices. Proving Turán's theorem would just require us to optimize this process for a maximal number of edges.

As before, our graph $K_{n_1,\ldots,n_{k-1}}$ will have a maximal number of edges when the n_i are balanced so that the subsets all have the same number of vertices, or at worst are within 1 of each other when the total number of vertices is not evenly divisible by $k-1$. For if our subset sizes were unbalanced, that is, if $n_i - n_j \geq 2$ for some V_i and V_j, we can switch a vertex from V_i to V_j. Then the number of edges between V_i and V_j changes from $n_i n_j$ to

$$(n_i - 1)(n_j + 1) = n_i n_j + n_i - n_j - 1 > n_i n_j.$$

We also have that the number of edges between V_i and V_j with any other set does not change. So, heuristically, the optimal graphs for this approach are $K_{n_1,\ldots,n_{k-1}}$ where $|n_i - n_j| \leq 1$ for all i, j; these are called **Turán graphs**.

In particular, if we can equally distribute the N vertices (that is, N is divisible by $k-1$), we get the Turán graph $K_{(n,\ldots,n)}$ where $n = \frac{N}{k-1}$. The number of edges in this graph is

$$\binom{k-1}{2} n^2 = \frac{(k-1)(k-2)}{2} \frac{N^2}{(k-1)^2} = \left(1 - \frac{1}{k-1}\right) \frac{N^2}{2}.$$

This value is known as the $(k-1)$st *Turán number*, $t_{k-1}(N)$. In 1941, Pál Turán proved that these graphs do in fact provide the best bound possible. Next, we give a formal proof of this result.

Proof of Turán's theorem. Let $G = (V, E)$ be a graph on N vertices which does not have a k-clique. We are going to transform G into a graph H which has at least as many edges as G and still does

1.5. Turán's theorem

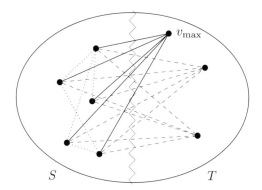

Figure 1.14. Constructing the graph H: The edges between nodes in S remain (dotted); the edges between nodes in T are removed; every vertex in S is connected to every vertex in T (dashed).

not have a k-clique in the following manner: Choose a vertex $v_{\max}$ of maximal degree. We can now partition our vertex set into two subsets; let S be the subset of vertices in G adjacent to $v_{\max}$ and let $T := V \setminus S$. To go from G to H, first remove any edges between vertices in T. Then, for every vertex $v \in T$ and vertex $v' \in S$, connect v to v' (if they are not already connected). See Figure 1.14. Note that $v_{\max} \in T$.

We will now demonstrate that the number of edges in H is no smaller than the number of edges in G. To do so, we will utilize the fact that

$$|E| = \frac{1}{2} \sum_{v \in V} d(v),$$

where $d(v)$ is the degree of the vertex v. It is enough to show that $d_H(v) \geq d_G(v)$ for every $v \in V$, i.e. every vertex has degree at least as high in H as in G.

If $v \in T$, then, by our construction, $d_H(v) = d_G(v_{\max}) \geq d_G(v)$ since v_m had maximal degree.

If $v \in S$, we know the degree of v in H can only increase since it is adjacent to the same vertices in S and now also adjacent to every vertex in T.

We claim that H has no k-clique. Clearly, any clique can have at most one vertex in T and so it suffices to show that S does not have a $(k-1)$-clique. However, this is true since G has no k-cliques: If we had a $(k-1)$-clique in S, we could add the vertex $v_{\max}$, which would be adjacent to every vertex in the clique, thus forming a k-clique in G.

We can now apply the same transformation to the subgraph induced by S, and inductively to the corresponding version of S in that graph. In this way, we eventually end up with a $(k-1)$-partite graph. (Note that, by construction, none of the vertices in T share an edge in H.) And our construction also shows that if G is a graph on N vertices with no k-clique, there is also a $(k-1)$-partite graph on N vertices that has at least as many edges as G. But we already know that among the $(k-1)$-partite graphs, the Turán graphs have the maximal number of edges. □

1.6. The finite Ramsey theorem

Two-element subsets of a set S can be represented as graphs, and this provided a visual framework for much of this chapter. Ramsey's theorem, however, holds not only for pairs, but in general for arbitrary p-element subsets.

Theorem 1.31 (Ramsey's theorem in its general form). *For any $r, k \geq 2$ and $p \geq 1$, there exists some integer N such that*

$$N \longrightarrow (k)_r^p.$$

One can represent p-element subsets as *hypergraphs*. In a hypergraph, any (non-zero) number of vertices can be joined by an edge, instead of just two (as in graphs). In other words, for a hypergraph, the edge set is a subset of $\mathcal{P}(V) \setminus \{\varnothing\}$, where $\mathcal{P}(V)$ is the *power set* of V. If the number of vertices in an edge is constant throughout the hypergraph, we speak of a *uniform* hypergraph. The hypergraphs of interest in the general Ramsey theorem are therefore p-uniform hypergraphs. For example, in Figure 1.15 we see 3-uniform hypergraphs.

Recall that when we first proved Ramsey's theorem for graphs (the $p = 2$ case of the theorem above), we relied heavily on the pigeonhole principle (the $p = 1$ case). This suggests that we might want

1.6. The finite Ramsey theorem

to try to use induction on p. With this in mind, let's see how we can prove the $p = 3$ case from what we already know, and then show how to proceed with the induction. We will also focus on the case of two colors ($r = 2$) for now and then discuss what needs to be changed to adapt our argument for more than two colors.

A proof for $p = 3$ and $r = 2$. The case of $p = 3$ and $r = 2$ is just "one step up" from Ramsey's theorem for graphs, Theorem 1.16. Instead of pairs, we are coloring *triples*.

So let us suppose that a coloring

$$c : [N]^3 \to \{\text{red, blue}\}$$

is given, where as before we imagine N to be a sufficiently large number for now.

We can try to emulate the proof of Theorem 1.16. There, we started by picking a arbitrary vertex v_1 and partitioned the remaining vertices according to the color of the edge connecting each vertex to v_1. This suggests starting by picking *two* numbers, say a_1 and a_2, from $\{1,\ldots,N\}$. Let $S_2 := \{1,\ldots,N\} \smallsetminus \{a_1, a_2\}$. We can partition S_2 into two sets: one set S_2^{red} containing those elements $x \in S_2$ such that $\{a_1, a_2, x\}$ is colored red and the other set S_2^{blue} where the corresponding set is colored blue (see left diagram of Figure 1.15). This directly corresponds to the proof of Theorem 1.16. We pick starting elements and color the remaining ones depending on what color the triple has that they form with the starting elements.

In the proof of Theorem 1.16, we then restricted ourselves to the larger of the two subsets (colors). We can do this again. Call this larger set S_3. Without loss of generality, say that S_3 is the set of x where $\{a_1, a_2, x\}$ is colored red.

Now, choose $a_3 \in S_3$ and let $S_3' := S_3 \smallsetminus \{a_3\}$. In the proof of Theorem 1.16, we next looked at the colors of the edges connecting a_3 to the nodes in S_3'. Now, however, we have colored triples. How should we divide the numbers in S_3'?

We know that for all $x \in S_3'$ we have that $\{a_1, a_2, x\}$ is red, but $\{a_1, a_3, x\}$ and $\{a_2, a_3, x\}$ can be red or blue. The idea is to partition S_3' into *four* sets: the x such that both $\{a_1, a_3, x\}$ and $\{a_2, a_3, x\}$

1. Graph Ramsey theory

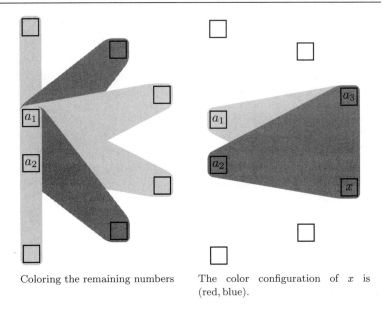

Coloring the remaining numbers   The color configuration of x is (red, blue).

Figure 1.15. The proof of Ramsey's theorem for triples

are red, the x such that both triples are blue, the x where the first triple is red and the second is blue, and the x where the first triple is blue and the second is red. Each set represents how x is colored with respect to the numbers already selected. Think of these sets as "*coloring configurations*" (see right diagram of Figure 1.15).

One of these four sets must be the largest (or, at least, not smaller than any of the other three); call this largest set S_4. Note that if i and j are distinct elements from $\{1, 2, 3\}$, then $c(a_i, a_j, x)$ is constant for all $x \in S_4$ and therefore depends not on our choice of x but only on a_i and a_j.

We can then continue inductively to find a sequence of elements $a_1, a_2, \ldots, a_t$: At the beginning of stage t, we have already defined a set $S_{t-1} = \{a_1, a_2, \ldots, a_{t-1}\}$. We first pick an arbitrary element a_t in S_{t-1} and put $S'_{t-1} = S_{t-1} \setminus \{a_t\}$. Next, we determine the color configurations of all $x \in S'_{t-1}$. This involves checking the color of all triples $\{a_i, a_t, x\}$ where $i < t$. Each color configuration can be thought of as a sequence of length $t-1$ recording the colors of the $\{a_i, a_t, x\}$, for

1.6. The finite Ramsey theorem

example (red, blue, red, red, ...). We see that there are 2^{t-1} possible color configurations for x. Now partition S'_{t-1} into 2^{t-1} sets, where x and y are in the same set if they have the same color configuration. From these 2^{t-1} sets, pick one of maximal size and make this set S_t.

As before, we have that if $1 \le i < j \le t$, then $c(a_i, a_j, x)$ is constant for all $x \in S_t$. The color depends only on our choice of i and j.

We can carry out this construction till we run out of numbers, that is, till S_t consists of one number only; call this number a_{t+1}.

We have constructed a sequence $a_1, a_2, \ldots, a_{t+1}$. This sequence does not yet define a monochromatic subset, but by the way we constructed it we can trim it down to one. The crucial property this sequence inherits from our construction is the following:

(1.4) Whenever we pick $1 \le i < j < s \le t+1$,
$c(a_i, a_j, a_s)$ depends only on i and j, not on s.

The reason is simple: a_s is eventually chosen from S_{s-1}, which is a subset of S_j, and for all numbers $x \in S_j$, we know they have the same color configuration when forming triples with numbers from $\{a_1, \ldots, a_j\}$. In particular, $\{a_i, a_j, x\}$ has the same color for all $x \in S_{j+1}$.

In the proof of Theorem 1.16, we used the pigeonhole principle in Stage 2 to trim the sequence of a_n down to a monochromatic one. Now we use Ramsey's theorem for graphs (i.e. this theorem for $p = 2$) to do so.

We define a new coloring of *pairs* from the sequence $(a_1, \ldots, a_t)$:

$$c^* : [\mathbb{N}]^2 \to \{\text{red}, \text{blue}\}$$
$$c^*(a_i, a_j) := c(a_i, a_j, a_s),$$

where $i < j < s \le t+1$. This coloring is well-defined by virtue of (1.4). Now, by Ramsey's theorem for graphs, if t is large enough so that $t \longrightarrow (k)_2^2$, we have a monochromatic subset $\{b_1, \ldots, b_k\} \subseteq \{a_1, \ldots, a_t\}$ for c^*. By our definition of c^*, this implies the existence of a monochromatic subset for c, because for any $\{b_{j_1}, b_{j_2}, b_{j_3}\}$ with $j_1 < j_2 < j_3$ we know that

$$c(b_{j_1}, b_{j_2}, b_{j_3}) = c^*(b_{j_1}, b_{j_2})$$

is constant.

To complete the proof, we have to argue that t can indeed be obtained large enough so $t \to (k)_2^2$. This is just a matter of picking the right *starting size* N. At each stage s of the construction of the a_i, we restrict ourselves to the largest set of remaining numbers with equal color configurations. As we argued above, there are 2^{s-1} possible color configurations, so if we have N_{s-1} numbers left entering stage s, we have at least

$$\frac{N_{s-1} - 1}{2^{s-1}}$$

numbers left over after the completion of stage s. If we choose N large enough, then the sequence

$$N, \frac{N-2}{2}, \frac{\frac{N-2}{2} - 1}{4}, \frac{\frac{\frac{N-2}{2}-1}{4} - 1}{8}, \ldots$$

reaches 1 after more than $R(k)$ many steps, where $R(k) = R(k,k)$ is the kth diagonal Ramsey number (see Definition 1.21). Therefore if N is chosen so that

$$N \geq 2^{(2R(k))^2},$$

a crude estimate, we obtain a monochromatic subset of size k.

From triples to p-tuples. It should now be clear how to lift this proof from $p = 3$ to arbitrary p, using induction. We construct a sequence of numbers $a_1, a_2, a_3, \ldots, a_{t+1}$ such that

($*_p$) whenever we pick $1 \leq i_1 < i_2 < \cdots < i_{p-1} < s \leq t+1$, the color of $\{a_{i_1}, a_{i_2}, \ldots, a_{i_{p-1}}, a_s\}$ depends only on $i_1, i_2, \ldots, i_{p-1}$, not on s.

The construction of such a sequence proceeds similarly to the case of $p = 3$, except that at stage t we look at the color configuration of p-sets $\{a_{i_1}, a_{i_2}, \ldots, a_{i_{p-1}}, x\}$.

We also define a coloring c^* as

$$c^* : [\mathbb{N}]^{p-1} \to \{\text{red}, \text{blue}\}$$
$$c^*(a_{i_1}, a_{i_2}, \ldots, a_{i_{p-1}}) := c(a_{i_1}, a_{i_2}, \ldots, a_{i_{p-1}}, a_s),$$

where c^* is well-defined by virtue of ($*_p$). The inductive hypothesis is that Ramsey's theorem holds for $p - 1$ and $r = 2$, and this gives

1.6. The finite Ramsey theorem

us, provided t is large enough so that $t \to (k)_2^{p-1}$, a monochromatic subset for c.

Again, an argument similar to that in the $p = 3$ case yields that N can indeed be chosen sufficiently large. But if we go over this calculation to find an estimate for N, we will see that the bookkeeping becomes even harder.

A different proof. We finish this section by giving a complete proof for the general case of p-sets and r-coloring. It is due to J. Nešetřil [46]. We denote by $R(p, k, r)$ the least natural number N such that $N \longrightarrow (k)_r^p$. Thus, $R(k) = R(2, k, 2)$. The claim is that $R(p, k, r)$ exists for all $k, p, r \geq 1$.

Call an r-coloring $c : [X]^p \to \{0, \ldots, r-1\}$ *good* if for any $x, y \in [X]^p$, if $\min x = \min y$, then $c(x) = c(y)$; that is, if two p-sets have the *same beginning*, then they have the same color. (We will encounter colorings of this type again in Section 4.6.)

Claim I: If $c : [r(k-1)+1]^p \to \{0, \ldots, r-1\}$ is good, then there exists $H \subseteq [r(k-1)+1]$ of size k such that c is monochromatic on $[H]^p$.

To see this, define a coloring $c^* : [r(k-1)+1] \to \{0, \ldots, r-1\}$ by letting

$$c^*(j) = \text{the unique color of every } p\text{-set with beginning } j$$

if $j \leq r(k-1) + 1 - p$ and $c^*(j) = 0$ otherwise.

By the pigeonhole principle, there exists a set $Y \subset [r(k-1)+1]$ of size k such that c^* is monochromatic on Y. But c^* being monochromatic implies that c is monochromatic on $[Y]^p$, since c is good.

By Claim I, it suffices to show that for any $p, k, r \geq 1$ there exists an N such that whenever c is an r-coloring of $[N]^p$, there exists a $H \subseteq [N]$ such that c is good on $[H]^p$ (since this implies that $R(p, k, r)$ exists and is at most $r(N-1) + 1$).

We use the following arrow notation for such an N:

$$N \xrightarrow{\text{good}} (k)_r^p.$$

Claim II: For any $p, k, r \geq 1$, there exists N with
$$N \xrightarrow{\text{good}} (k)_r^p.$$

One proves this claim by *double induction* on p and k. The inductive hypothesis is that for *all* k, there exists N with
$$N \xrightarrow{\text{good}} (k)_r^{p-1}.$$
Furthermore, the inductive hypothesis also assumes there exists N such that
$$N \xrightarrow{\text{good}} (k)_r^p.$$
We then show that there exists an N such that
$$N \xrightarrow{\text{good}} (k+1)_r^p.$$

Note that Claim I is verified independently and without induction, so we can assume not only that for all k there exists N with $N \xrightarrow{\text{good}} (k)_r^{p-1}$, but also that $R(p-1, k, r)$ exists.

Assume $N \xrightarrow{\text{good}} (k)_r^p$. We want to find M such that
$$M \xrightarrow{\text{good}} (k+1)_r^p.$$
We claim that $M = 1 + R(p-1, N, r)$ suffices. Namely, suppose $c: [M]^p \to \{0, \ldots, r-1\}$. Consider the coloring c' of $[\{2, \ldots, M\}]^{p-1}$ given by
$$c'(b_1, \ldots, b_{p-1}) = c(1, b_1, \ldots, b_{p-1}).$$
By the choice of M there exists a monochromatic subset Y of $\{2, \ldots, M\}$ of size N. By the definition of c', all p-sets in $[M]$ containing 1 have the same c-color. One can say that the coloring is good on 1. Now we have to refine the set Y to make it a good coloring overall.

But we know that $N \xrightarrow{\text{good}} (k)_r^p$ and $|Y| = N$, so we can find $Z \subset Y$ with $|Z| = k$, such that c is good on Z. Then c is also good on $\{1\} \cup Z$, a set of size $k+1$, as desired.

Chapter 2

Infinite Ramsey theory

2.1. The infinite Ramsey theorem

In this chapter, we will look at Ramsey's theorem for colorings of infinite sets. We start with the simplest infinite Ramsey theorem. We carry over the notation from the finite case. Given any set Z and a natural number $p \geq 1$, $[Z]^p$ denotes the set of all p-element subsets of Z, or simply the *p-sets of Z*.

Theorem 2.1 (Infinite Ramsey theorem). *Let Z be an infinite set. For any $p \geq 1$ and $r \geq 1$, if $[Z]^p$ is colored with r colors, then there exists an infinite set $H \subseteq Z$ such that $[H]^p$ is monochromatic.*

Compared with the finite versions of Ramsey's theorem in Chapter 1, the statement of the theorem seems rather elegant. This is due to a robustness of infinity when it comes to subsets: It is possible to remove infinitely many elements from an infinite set and still have an infinite set. It is customary to call a monochromatic set H as in Theorem 2.1 a **homogeneous** (for c) subset, and from here on we will use *monochromatic* and *homogeneous* interchangeably.

Proof. Fix $r \geq 1$. We will proceed via induction on p. For $p = 1$ the statement is the simplest version of an **infinite pigeonhole principle**:

If we distribute infinitely many objects into finitely many drawers, one drawer must contain infinitely many objects. In our case, the drawers are the colors $1, \ldots, r$, and the objects are the elements of Z.

Next assume $p > 1$ and let $c : [Z]^p \to \{1, \ldots, r\}$ be an r-coloring of the p-element subsets of Z.

To use the induction hypothesis, we fix an arbitrary element $z_0 \in Z$ and use c to define a coloring of $(p-1)$-sets: For $\{b_1, \ldots, b_{p-1}\} \in [Z \smallsetminus \{z_0\}]^{p-1}$, define

$$c_0(b_1, \ldots, b_{p-1}) := c(z_0, b_1, \ldots, b_{p-1}).$$

Note that $Z \smallsetminus \{z_0\}$ is still infinite. Hence, by the inductive hypothesis, there exists an infinite homogeneous $Z_1 \subseteq Z \smallsetminus \{z_0\}$ for c_0, which in turn means that all p-sets $\{z_0, b_1, \ldots, b_{p-1}\}$ with $b_1, \ldots, b_{p-1} \in Z_1$ have the same c-color.

Pick an element z_1 of Z_1. Now define a coloring of the $(p-1)$-sets of $Z_1 \smallsetminus \{z_1\}$: For $b_1, \ldots, b_{p-1} \in Z_1 \smallsetminus \{z_1\}$, put

$$c_1(b_1, \ldots, b_{p-1}) := c(z_1, b_1, \ldots, b_{p-1}).$$

Again, our inductive hypothesis tells us that there is an infinite homogeneous subset $Z_2 \subseteq Z_1 \smallsetminus \{z_1\}$ for c_1.

We can continue this construction inductively and obtain infinite sets $Z \supset Z_1 \supset Z_2 \supset Z_3 \supset \cdots$, where Z_{i+1} is homogeneous for a coloring c_i of the $(p-1)$-sets of Z_i that is derived from c by fixing one element z_i of Z_i, and thus all p-sets of $\{z_i\} \cup Z_{i+1}$ that contain z_i have the same c_i-color.

By virtue of our choice of the Z_i and the z_i, the sequence of the z_i has the crucial property that for any $i \geq 0$,

$$\{z_{i+1}, z_{i+2}, \ldots\}$$

is homogeneous for c_i (namely, it is a subset of Z_{i+1}). Let k_i denote the color ($\in \{1, \ldots, r\}$) for which the homogeneous set Z_{i+1} is monochromatic.

Now use the infinite pigeonhole principle one more time: At least one color, say k^*, must occur infinitely often among the k_i. Collect

2.2. König's lemma and compactness

the corresponding z_i's in a set H. We claim that H is homogeneous for c.

To verify the claim, let $\{h_1, h_2, \ldots, h_p\} \subset H$. Every element of H is a z_i, i.e. there exist $i_1, \ldots, i_p$ such that

$$h_1 = z_{i_1}, \ldots, h_p = z_{i_p}.$$

Without loss of generality, we can assume that $i_1 < i_2 < \cdots < i_p$ (otherwise reorder). Then $\{h_1, h_2, \ldots, h_p\} \subset Z_{i_1}$, and hence the color of $\{h_1, h_2, \ldots, h_p\}$ is $k_{i_1} = k^*$ (all the colors corresponding to a z_j in H are equal to k^*). The choice of $\{h_1, h_2, \ldots, h_p\}$ was arbitrary, and thus H is homogeneous for c. □

Conceptually, this proof is not really different from the proofs of the finite Ramsey theorem, Theorem 1.31. We start with an arbitrary element of Z and "thin out" the set $\mathbb{N}$ so that all possible completions of this element to a p-set have the same c-color. We pick one of the remaining elements and do the same for all other remaining elements, and so on. Then we apply the pigeonhole principle one more time to homogenize the colors. The difference is that in the finite case we argued that if we start with enough numbers (or vertices), the process will produce a large enough finite sequence of numbers (vertices). In the infinite case, the process *never stops*. This is the robustness of infinity mentioned above: It is possible to take out infinitely many elements infinitely many times from an infinite set and still end up with an infinite set. In some sense, the set we end up with (H) is smaller than the set we started with (Z). But in another sense, it is of the same size: it is still infinite.

This touches on the important concept of *infinite cardinalities*, to which we will return in Section 2.5.

2.2. König's lemma and compactness

As noted before, the infinite Ramsey theorem is quite elegant, in that its nature seems *more qualitative than quantitative*. We do not have to worry about keeping count of finite cardinalities. Instead, the robustness of infinity takes care of everything.

It is possible to exploit infinitary results to prove finite ones. This technique is usually referred to as **compactness**. The essential ingredient is a result about infinite trees known as **König's lemma**. This is a purely combinatorial statement, but we will see in the next section that it can in fact be seen as a result in *topology*, where compactness is originally rooted.

Using compactness relieves us of much of the counting and bookkeeping we did in Chapter 1, but usually at the price of not being able to derive bounds on the finite Ramsey numbers. In fact, using compactness often introduces huge numbers. In Chapters 3 and 4, we will see how large these numbers actually get.

Partially ordered sets. In Section 1.2, we introduced trees as a special family of graphs (those without cycles). We also saw that every tree induces a *partial order* on its vertex set. Conversely, if a partial order satisfies certain additional requirements, it induces a tree structure on its elements, on which we will now elaborate.

Orders play a fundamental role not only in mathematics but in many other fields from science to finance. Many things we deal with in our life come with some characteristics that allow us to compare and order them: gas mileage or horse power in cars, interest rates for mortgages, temperatures in weather reports—the list of examples is endless. Likewise, the mathematical theory of orders studies sets that come equipped with a binary relation on the elements, the *order*.

Most mathematical orders you encounter early on are *linear*, and they are so natural that we often do not even realize there is an additional structure present. The integers, the rationals, and the reals are all linearly ordered: If we pick any two numbers from these sets one will be *smaller* than the other. But we can think of examples where this is not necessarily the case. For example, take the set $\{1, 2, 3\}$ and consider all possible subsets:

$$\varnothing, \{1\}, \{2\}, \{3\}, \{1,2\}, \{1,3\}, \{2,3\}, \{1,2,3\}.$$

We order these subsets by saying that A is smaller than B if $A \subset B$. Then $\{1\}$ is smaller than $\{1, 2\}$, but what about $\{1, 2\}$ and $\{1, 3\}$? Neither is contained in the other, and so the two sets are *incomparable* with respect to our order—the order is *partial* (Figure 2.1).

2.2. König's lemma and compactness

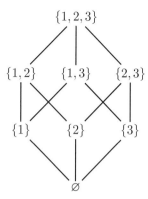

Figure 2.1. The subset partial order on $\{1,2,3\}$

The notion of a **partially ordered set** captures the minimum requirements for a binary relation on a set to be meaningfully considered a partial order.

Definition 2.2. Let X be a set. A **partial order** on X is a binary relation $<$ on X such that

(P1) for all $x \in X$, $x \not< x$ (*irreflexive*);

(P2) for all $x, y, z \in X$, if $x < y$ and $y < z$, then $x < z$ (*transitive*).

The pair $(X, <)$ is often simply called a **poset**. A partial order $<$ on X is **linear** (also called *total*) if additionally

(L) for all $x, y \in X$, $x < y$ or $x = y$ or $y < x$.

If $(X, <)$ is a poset one writes, as usual, $x \leq y$ to express that either $x < y$ or $x = y$. As we saw above, the usual order on the integers, rationals, and reals is linear, while the subset-ordering of the subsets of $\{1,2,3\}$ is a partial order but not linear. Another example of a partial order that is not linear is the following: Let $X = \mathbb{R}^2$, and for $x = (x_1, x_2)$ and $y = (y_1, y_2)$ in $\mathbb{R}^2$ put

$$x < y \quad \Longleftrightarrow \quad \|x\| < \|y\|,$$

that is, we order vectors by their length. This order is not linear since for each length $l > 0$, there are infinitely many vectors of length l (which therefore cannot be compared).

Trees from partial orders. Let $(T, <)$ be a partially ordered set. $(T, <)$ is called a **tree** (as a partial order) if

(T1) there exists an $r \in T$ such that for all $x \in T$, $r \le x$
(r is the *root* of the tree);

(T2) for any $x \in T$, the set of *predecessors* of x, $\{y \in T: y < x\}$, is finite and linearly ordered by $<$.

Note that not every poset is a tree. For example, in the set of all subsets of $\{1, 2, 3\}$, the predecessors of $\{1, 2, 3\}$ are not linearly ordered. Often a poset also lacks a root element. For example, the usual ordering of the integers $\mathbb{Z}$, $\cdots < -2 < -1 < 0 < 1 < 2 < \cdots$, satisfies neither (T1) nor (T2).

Trees arising from partial orders can be interpreted as graph-theoretic trees, as introduced in Section 1.2. In fact, the elements of T are called *nodes*, and sets of the form $\{y \in T: y \le x\}$ are called *branches*.

Exercise 2.3. Let $(T, <)$ be a tree (partial order). Define a graph by letting the node set be T and connect two nodes if one is an *immediate predecessor* of the other. (Node s is an immediate predecessor of t if $s < t$ and if for all $u \in T$, $u < t$ implies $u \le s$.) Show that the resulting graph is a tree in the graph-theoretic sense.

As an example, consider the set $\{0, 1\}^*$ of all *binary strings*. A binary string σ is a finite sequence of 0s and 1s, for example,

$$\sigma = 01100010101.$$

We order strings via the *initial segment relation*: $\sigma < \tau$ if σ is shorter than τ and the two strings agree on the bits of σ. For example, 011 is an initial segment of 01100, but 010 is not (the two strings disagree on the third bit). It is not hard to verify that (P1) and (P2) hold for this relation. Furthermore, the *empty string* λ is an initial segment of any other string and the initial segments of a string are linearly ordered by $<$; for example, for $\sigma = 010010$,

$$\lambda < 0 < 01 < 010 < 0100 < 01001 < \sigma.$$

Therefore, $(\{0, 1\}^*, <)$ is a tree, the **full binary tree** (Figure 2.2).

2.2. König's lemma and compactness   47

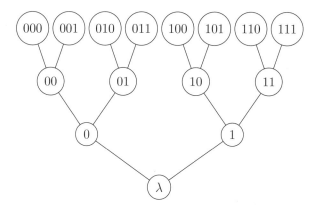

Figure 2.2. The full binary tree up to strings of length 3

Paths and König's lemma. While we think of branches as something finite—due to (T2)—a sequence of them can give rise to an infinite branch, also called an *infinite path*.

Definition 2.4. An **infinite path** in a tree $(T, \leq)$ is a sequence $r = x_0 < x_1 < x_2 < \cdots$ where all $x_i \in T$ and for all i, $\{x \in T : x < x_i\} = \{x_0, x_1, \ldots, x_{i-1}\}$.

One can think of an infinite path as a sequence of elements on the tree where each element is a "one-step" extension of the previous one. In the full binary tree, an infinite path corresponds to an infinite sequence of zeros and ones.

It is clear that if a tree T has a path, then it must be infinite (i.e. T as a set is infinite). But does the converse hold? That is:

If a tree is infinite, does it have a path?

It is easy to give an example to show that this is not true. Consider the tree on $\mathbb{N}$ where 0 is the root and every number $n \geq 1$ is an immediate successor of 0 (an infinite fan of depth one).

Definition 2.5. A tree $(T, \leq)$ is *finitely branching* if for every $x \in T$, there exist at most finitely many $y_1, \ldots, y_n \in T$ such that whenever $z > x$ for some $z \in T$, we have $z \geq y_i$ for some i. (That is, every $x \in T$ has at most finitely many immediate successors.)

Our example of the full binary tree $\{0,1\}^*$ is finitely branching.

Theorem 2.6 (König's lemma). *If an infinite tree $(T, <)$ is finitely branching, then it has an infinite path.*

Proof. We construct an infinite sequence on T by induction. Let $x_0 = r$. Given any $x \in T$, let T_x denote the part of T "above" x, i.e.

$$T_x = \{y \in T \colon x \le y\}.$$

T_x inherits a tree structure from T by letting x be its root. Note that $T = T_r$. Let $y_1^{(r)}, \ldots, y_n^{(r)}$ be the immediate successors of r in T. Since we assume T to be infinite, by the infinite pigeonhole principle, one of the trees $T_{y_1^{(r)}}, \ldots, T_{y_n^{(r)}}$ must be infinite, say $T_{y_i^{(r)}}$. Put $x_1 = y_i^{(r)}$.

We can now iterate this construction, using the infinite pigeonhole principle on the finitely many disjoint trees above the immediate successors of x_1, and so on. Since we always maintain an infinite tree above our current x_k, the construction will carry on indefinitely and we obtain an infinite sequence $x_0 < x_1 < x_2 < \cdots$, an infinite path through T. □

Proving finite results from infinite ones. We can use König's lemma and the infinite Ramsey theorem (Theorem 2.1) to prove the general finite Ramsey theorem (Theorem 1.31).

Assume, for the sake of contradiction, that for some k, p, and r, the statement of the finite Ramsey theorem does not hold. That is, for all n there exists at least one coloring $c_n \colon [n]^p \to 1, \ldots, r$ such that no monochromatic subsets of size k exist. Collect these counterexamples c_n, for all n, in a single set T, and order them by *extension*: Let $c_m < c_n$ if and only if $m < n$ and the restriction of c_n to $[m]^p$ is equal to c_m, i.e. c_m extends c_n as a function.

We make three crucial observations:

(1) $(T, <)$ is a tree; the root r is the empty function and the predecessors of a coloring $c_n \in T$ are the restricted colorings

$$\varnothing < c_n|_{[p]^p} < c_n|_{[p+1]^p} < \cdots < c_n|_{[n-1]^p} < c_n.$$

(2) T is finitely branching; this is clear since for every n there are only finitely many functions $c \colon [n]^p \to \{1, \ldots, r\}$ at all.

2.2. König's lemma and compactness

(3) T is infinite; this is true because we are assuming that at least one such coloring exists for all n.

Therefore, we can apply König's lemma and obtain an infinite path
$$\varnothing < c_p < c_{p+1} < c_{p+2} < \cdots.$$
Since each c_n on our path is an extension of all the previous functions on the path, we can construct a well-defined function $C : [\mathbb{N}]^p \to \{1,\ldots,r\}$ which has the property that $C|_{[n]^p} = c_n$. That is, if X is a p-set of $\mathbb{N}$ where N is the largest integer in X, then $C(X) = c_N(X)$.

Now we have a coloring on $\mathbb{N}$ and we can apply the infinite Ramsey theorem to deduce that there exists an infinite subset $H \subseteq \mathbb{N}$ such that $[H]^p$ is monochromatic.

We write $H = \{h_1 < h_2 < h_3 < \cdots\}$ and let $N = h_k$. Since H is monochromatic for the coloring C, so is every subset of H. In particular, $H_k := \{h_1 < h_2 < \cdots < h_k\}$ is a monochromatic subset of size k for the coloring C. But we have $C(H_k) = c_N(H_k)$, and therefore c_N has a monochromatic subset of size k, which contradicts our initial assumption.

A blueprint for compactness arguments. We can use the previous proof as a prototype for future uses of compactness. Suppose we have a statement $P(\vec{\pi})$ with a vector of parameters $\vec{\pi}$ that asserts the existence of a certain object, and we want to show that for a sufficiently large finite set $\{1,\ldots,N\}$, $P(\vec{\pi})$ is always true. Suppose further that we have shown $P(\vec{\pi})$ is true for $\mathbb{N}$. Here is a blueprint for an argument using compactness:

(1) Assume, for a contradiction, that the finite version of P fails.

(2) Then we can find counterexamples for every set $[n]$.

(3) Collect these counterexamples in a set T, order them by extension, and show that under this ordering T forms an infinite, finitely branching tree.

(4) Apply König's lemma to obtain an infinite path in T, which corresponds to an instance of our statement $P(\vec{\pi})$ for $\mathbb{N}$.

(5) Since $P(\vec{\pi})$ is true for $\mathbb{N}$, we can choose a witness example for this instance.

(6) By restricting the witness to a sufficiently large subset, we obtain a contradiction to the fact that T contains only counterexamples to P.

We will later see that the introduction of infinitary methods opens a fascinating metamathematical door: There exist "finitary" statements (in the sense that all objects involved—sets, functions, numbers–are finite) for which *only infinitary proofs exist*. In a certain sense, the finite sets whose existence the infinitary methods establish are so huge that a "finitary accounting method" cannot keep track of them. We will investigate this phenomenon in Chapter 4.

2.3. Some topology

If you have learned about compactness in a topology or analysis class, you might be wondering why we are using this word. We will show that König's lemma can be rephrased in terms of sequential compactness. While we will provide all the necessary definitions, we can of course not even scratch the surface of the theory of metric spaces and topology. For a reader who has no previous experience in this area, we recommend consulting on the side one of the numerous textbooks on analysis or topology, for example [**51**].

Metric spaces. The concept of a metric space is a generalization of distance. We use it to describe how close or far two elements in a set are from each other.

Definition 2.7. A **metric** on a set X is any function $d : X^2 \to \mathbb{R}$ such that for all $x, y, z \in X$:

(1) d is *non-negative*, that is, $d(x, y) \geq 0$, and moreover $d(x, y) = 0$ if and only if $x = y$;

(2) d is *symmetric*, that is, $d(x, y) = d(y, x)$; and

(3) d satisfies the *triangle inequality*, that is, $d(x, z) \leq d(x, y) + d(y, z)$.

A **metric space** (X, d) is a set X together with a metric d.

The prototypical examples of metric spaces include $\mathbb{R}$ with the standard distance $d(x, y) = |x - y|$, or $\mathbb{R}^2$ with the distance function

2.3. Some topology

which comes from the Pythagorean theorem,
$$d(x,y) = \sqrt{(x_1 - y_1)^2 + (x_2 - y_2)^2}.$$
These are both examples of the n-dimensional Euclidean metric: For $x, y \in \mathbb{R}^n$,
$$d(x,y) = \left(\sum_{i=1}^{n} (x_i - y_i)^2\right)^{1/2}.$$
This metric is the natural function with which we associate the idea of "distance" between two points. However, there are many other important metrics. For any non-empty set X, we can consider the *discrete metric*, defined by
$$d(x,y) = \begin{cases} 0 & \text{if } x = y \\ 1 & \text{if } x \neq y \end{cases}$$
This is a metric where every two distinct points are the same distance away—a rather crude measure of distance.[1] One can refine this idea to take the combinatorial structure into account. For any connected graph G, we can define a metric on G by letting
$$d(v, w) = \text{length of the shortest path between } v \text{ and } w.$$
Exercise 2.8. Show that $d(v, w)$ defines a metric on a connected graph G.

Neighborhoods and open sets. Given a point x in a metric space (X, d) and a real number ϵ, the ϵ-**neighborhood of** x is the set
$$B_\epsilon(x) := \{y \in X : d(x,y) < \epsilon\},$$
i.e. all points which are less than ϵ away from x.

For example, an open ball $B_\varepsilon(x)$ (with respect to the Euclidean metric) on the real line $\mathbb{R}$ is just an open interval of the form $(x - \varepsilon, x + \varepsilon)$.

An **open set** is any set $U \subseteq X$ where, for every x in U, there exists an $\epsilon > 0$ such that $B_\epsilon(x)$ is contained entirely in U. The union of open sets is also an open subset. The complement of an open set is called a **closed set**. Note that in any metric space (X, d), the entire set X and the empty set $\varnothing$ are both open and closed.

---
[1] The mathematician Stanislaw Ulam once wrote that Los Angeles is a discrete metric space, where the distance between any two points is an hour's drive [**67**].

We can now define the topological notion of compactness using open coverings: A collection of open subsets $\{U_i\}_{i \in I}$ is defined to be an *open cover* of Y if $Y \subseteq \bigcup U_i$.

Definition 2.9. A subset Y in a metric space (X, d) is **compact** if whenever $\{U_i\}_{i \in I}$ is an open cover of Y, there is a finite subset $J \subseteq I$ such that $\{U_j\}_{j \in J}$ is also an open cover of Y, in other words, if every open cover has a finite subcover.

Suppose we cover $\mathbb{R}$ with balls $B_\varepsilon(x)$ for an arbitrarily small ε, so that every $x \in \mathbb{R}$ contributes an interval $(x - \varepsilon, x + \varepsilon)$. Together these intervals clearly cover all of $\mathbb{R}$. It is easy to see that this covering has no finite subcover, as choosing finitely many of the $B_\varepsilon(x)$ covers at most an interval of the form $(-M, M)$, where $M < \infty$. Therefore, $\mathbb{R}$ with the Euclidean metric is not a compact space.

This example suggests that compact sets, although possibly infinite (as sets), should somehow be considered "finite". In Euclidean space this is confirmed by the **Heine-Borel theorem**: A set $X \subseteq \mathbb{R}^n$ is compact if and only if it is closed and bounded, that is, X is contained in some n-dimensional cube $(-M, M)^n$.

Sequential compactness. Given a sequence of points (x_i) in a metric space (X, d), the sequence **converges** to a point x if

$$\lim_{i \to \infty} d(x_i, x) = 0.$$

A metric space (X, d) is **sequentially compact** if every sequence has a convergent subsequence. In metric spaces, the notion of compactness and sequential compactness are equivalent (see [51]), although this is not true for general topological spaces.

In $\mathbb{R}^n$ (with the Euclidean metric) the equivalence of compactness and sequential compactness follows from the *Bolzano-Weierstrass theorem*: Every bounded sequence has a convergent subsequence.

Exercise 2.10. Use the infinite Ramsey theorem to prove that every sequence in $\mathbb{R}$ has a monotone subsequence. As any bounded, monotone sequence converges in $\mathbb{R}$, this implies the Bolzano-Weierstrass theorem.

2.3. Some topology

Infinite trees as metric spaces. Given an infinite, finitely branching tree T, König's lemma tells us that there will be at least one infinite path. We can collect all of the infinite paths into a set, denoted by $[T]$. It will be useful to visualize the elements of $[T]$ not just as paths on a tree, but also as infinitely long sequences of nodes.

Suppose we have two elements of $[T]$, $\vec{s}$ and $\vec{t}$, and their sequences of nodes

$$\vec{s} = \{r = s_0 < s_1 < s_2 < \cdots\} \text{ and } \vec{t} = \{r = t_0 < t_1 < t_2 < \cdots\}.$$

To define a notion of distance, two paths will be regarded as "close" if their sequences agree for a long time. We put, for distinct $\vec{s}$ and $\vec{t}$,

$$D_{\vec{s},\vec{t}} = \min\{i \geq 0 \colon s_i \neq t_i\}$$

and then define our distance function as

$$d(\vec{s}, \vec{t}) = \begin{cases} 0 & \text{if } s = t, \\ 2^{-D_{\vec{s},\vec{t}}} & \text{if } s \neq t. \end{cases}$$

We claim that d is a metric on $[T]$ and will call it the **path metric** on $[T]$. Non-negativity and symmetry are clear from the definition of d. To verify the triangle inequality, suppose $\vec{s}, \vec{t},$ and $\vec{u}$ are pairwise distinct paths in $[T]$. (If any two of the sequences are identical, the statement is easy to verify.) We distinguish two cases:

Case 1: $D_{\vec{s},\vec{u}} \leq D_{\vec{s},\vec{t}}$.

This means that $\vec{s}$ agrees at least as long with $\vec{t}$ as with $\vec{u}$. But this implies that $\vec{t}$ agrees with $\vec{u}$ precisely as long as $\vec{s}$ does, which in turn means $D_{\vec{t},\vec{u}} = D_{\vec{s},\vec{u}}$, and hence

$$d(\vec{s}, \vec{u}) = 2^{-D_{\vec{s},\vec{u}}} = 2^{-D_{\vec{t},\vec{u}}} \leq 2^{-D_{\vec{s},\vec{t}}} + 2^{-D_{\vec{t},\vec{u}}} = d(\vec{s}, \vec{t}) + d(\vec{t}, \vec{u}).$$

Case 2: $D_{\vec{s},\vec{u}} > D_{\vec{s},\vec{t}}$.

In this case $\vec{s}$ agrees with $\vec{u}$ longer than it agrees with $\vec{t}$. But this directly implies that

$$d(\vec{s}, \vec{u}) = 2^{-D_{\vec{s},\vec{u}}} < 2^{-D_{\vec{s},\vec{t}}} \leq 2^{-D_{\vec{s},\vec{t}}} + 2^{-D_{\vec{t},\vec{u}}} = d(\vec{s}, \vec{t}) + d(\vec{t}, \vec{u}).$$

What do the neighborhoods $B_\varepsilon(s)$ look like for this metric? A sequence t is in $B_\varepsilon(s)$ if and only if $2^{-D_{\vec{s},\vec{t}}} < \varepsilon$, which means $D_{\vec{s},\vec{t}} > -\log_2 \varepsilon$. Hence t is in the ε-neighborhood of $\vec{s}$ *if and only if it agrees with $\vec{s}$ on the first $\lceil -\log_2 \varepsilon \rceil$ bits.*

Exercise 2.11. Draw a picture of $B_{1/8}(10101010\ldots)$.

König's lemma and compactness. We can now interpret König's lemma as an instance of topological compactness.

Theorem 2.12. *If T is a finitely branching tree, then $[T]$ with the path metric is a compact metric space.*

Proof. Assume that $[T]$ is not empty. Let $(\vec{s}_n)$ be a sequence in $[T]$. (We use the vector notation, afterall.) This means that every $\vec{s}_n$ is itself a sequence $r = s_n^0 < s_n^1 < s_n^2 < \cdots$ in T. We will construct a convergent subsequence $(\vec{s}_{n_i})$. It follows that $[T]$ is sequentially compact and therefore compact.

Let T^* be the subtree of T defined as follows: Let $\sigma \in T^*$ if and only if $\sigma = r$ or $\sigma = s_n^i$ for some i and n; that is, σ is in T^* if and only if it occurs in one of the paths $\vec{s}_n$.

We observe that T^* is infinite, because each $\vec{s}_n$ is an infinite path. It is also finitely branching as it is a subtree of the finitely branching tree T.

By König's lemma, T^* has an infinite path $\vec{t}$ of the form $r = t^0 < t^1 < t^2 < t^3 < \cdots$. We use this path to identify a subsequence of $(\vec{s}_n)$ that converges to $\vec{t} \in [T]$.

The path $\vec{t}$ is built from nodes that occur as a node in *some* path $\vec{s}_n$. This means that for every i there exists an n such that $t^i = s_n^i$. We use this to define a subsequence of (s_n) as follows: Let

$$n_i = \min\{n : s_n^i = t^i\}.$$

Claim: $(\vec{s}_{n_i})$ converges to $\vec{t}$.

By the definition of the path metric d,

$$d(\vec{s}_{n_i}, \vec{t}) \xrightarrow{i \to \infty} 0 \quad \text{iff} \quad \vec{s}_{n_i} \text{ and } \vec{t} \text{ agree}$$

on longer and longer segments.

But this is built into the definition of $(\vec{s}_{n_i})$: We have

$$s_{n_i}^i = t^i,$$

and since initial segments in trees are unique, this implies that

$$s_{n_i}^0 = t^0, s_{n_i}^1 = t^1, \ldots, s_{n_i}^{i-1} = t^{i-1}.$$

Therefore, $\vec{s}_{n_i}$ and $\vec{t}$ have an agreement of length i, and thus

$$d(\vec{s}_{n_i}, \vec{t}) \xrightarrow{i \to \infty} 0.$$

□

Exercise 2.13. Every real number $x \in [0, 1]$ has a *dyadic expansion* $\vec{s}^{(x)} \in \{0, 1\}^\infty$ such that

$$x = \sum_i s_i^{(x)} 2^{-i}.$$

The expansion is unique except when x is of the form $m/2^n$, with $m, n \leq 1$ integer. To make it unique, we require the dyadic expansion to eventually be constant $\equiv 1$.

Show that a sequence (x_i) of real numbers in $[0, 1]$ converges with respect to the Euclidean metric if and only if $(\vec{s}^{(x_i)})$ converges with respect to the path metric d.

2.4. Ordinals, well-orderings, and the axiom of choice

The natural numbers form the mathematical structure that we use to count things. In the process of counting, we bestow an order on the objects we are counting. We speak of the first, the second, the third element, and so on. The realm of the natural numbers is sufficient as long as we count only finite objects. But how can we count infinite sets? This is made possible by the theory of *ordinal numbers*.

Properties of ordinals. Ordinal numbers are formally defined using set theory, as transitive sets that are well-ordered by the element relation $\in$. We will not introduce ordinals formally here, but instead simply list some crucial properties of ordinal numbers that let us extend the counting process into the infinite realm. For a formal development of ordinals, see for example [**35**].

(O1) Every natural number is an ordinal number.

(O2) The ordinal numbers are *linearly ordered*, and 0 is the least ordinal.

(O3) Every ordinal has a unique *successor* (the *next* number); that is, for every ordinal α there exists an ordinal $\beta > \alpha$ such that
$$(\forall \gamma)\, \alpha < \gamma \Rightarrow \beta \leq \gamma.$$
The successor of α is denoted by $\alpha + 1$.

(O4) For every set A of ordinals, there exists an ordinal β that is the *least upper bound* of A, that is,

for all $\alpha \in A$,

$\alpha \leq \beta$ and β is the least number with this property.

If we combine (O1) and (O4), there must exist a least ordinal that is *greater than every natural number*. This number is called ω. (O3) tells us that ω has a successor, $\omega + 1$, which in turn has a successor itself, $(\omega+1)+1$, which we write as $\omega+2$. We can continue this process and obtain
$$\omega,\ \omega + 1,\ \omega + 2,\ \omega + 3,\ \ldots,\ \omega + n,\ \ldots$$
But the ordinals do not stop here. Applying (O4) to the set $\{\omega+n : n \in \mathbb{N}\}$, we obtain a number that is greater than any of these, denoted by $\omega+\omega$. Here is a graphical representation of these first infinite ordinals:

$$
\begin{array}{ll}
\omega & \circ\ \circ\ \circ\ \cdots \\
\omega+1 & \circ\ \circ\ \circ\ \cdots\ \bullet \\
\omega+2 & \circ\ \circ\ \circ\ \cdots\ \bullet\ \circ \\
\omega+\omega & \circ\ \circ\ \circ\ \cdots\ \bullet\ \circ\ \circ\ \cdots
\end{array}
$$

One can continue enumerating:
$$\omega+\omega+1,\ \omega+\omega+2,\ \ldots,\ \omega+\omega+\omega,\ \ldots,\ \omega+\omega+\omega+\omega,\ \ldots$$
In this process we encounter two types of ordinals.

- *Successor ordinals*: Any ordinal α for which there exists an ordinal β such that $\alpha = \beta + 1$. Examples include all natural numbers greater than 0, $\omega + 1$, and $\omega + \omega + 3$.

2.4. Ordinals, well-orderings, and axiom of choice

- *Limit ordinals*: Any ordinal that is not a successor ordinal, for example 0, ω, and $\omega + \omega + \omega$.

Since there is always a successor, the process never stops. An attentive reader will remark that, on the other hand, we could apply (O4) to the set of *all* ordinals. Would this not yield a contradiction? This is known as the *Burali-Forti paradox*. We have to be careful which mathematical objects we consider a set and which not. And in our case we say:

There is no set of all ordinals.

There are simply *too many* to form a set. The ordinals form what is technically referred to as a *proper class*, which we will denote by Ord. Other examples of proper classes are the class of all sets and the class of all sets that do not contain themselves (this is *Russell's paradox*). The assumption that either of these is a set leads to a contradiction similar to assuming that a set of all ordinals exists. Classes behave in many ways like sets—for example, we can talk about the *elements* of a class. But these elements cannot be other classes; classes are *too large* to be an element of something.

Ordinal arithmetic. The counting process above indicates that we can define arithmetical operations on ordinals similar to the operations we have on the natural numbers. For the natural numbers, addition and multiplication are defined by *induction*, by means of the following identities.

- $m + (n+1) = (m+n) + 1$,
- $m \cdot (n+1) = m \cdot n + m$.

For ordinals, we use *transfinite induction*. This is essentially the same as "ordinary" induction, except that we also have to account for limit ordinals in the induction step.

Addition of ordinals.

$$\begin{aligned}
\alpha + 0 &= \alpha, \\
\alpha + (\beta + 1) &= (\alpha + \beta) + 1, \\
\alpha + \lambda &= \sup\{\alpha + \gamma : \gamma < \lambda\} \quad \text{if } \lambda \text{ is a limit ordinal.}
\end{aligned}$$

There is an important aspect in which ordinal addition behaves very differently from addition for natural numbers. Take, for example,

$$1 + \omega = \sup\{1 + n : n \in \mathbb{N}\} = \sup\{m : m \in \mathbb{N}\} = \omega \neq \omega + 1.$$

So ordinal addition is **not commutative**; that is, it is not true in general that $\alpha + \beta = \beta + \alpha$.

Multiplication of ordinals.

$$\begin{aligned}
\alpha \cdot 0 &= 0, \\
\alpha \cdot (\beta + 1) &= (\alpha \cdot \beta) + \alpha, \\
\alpha \cdot \lambda &= \sup\{\alpha \cdot \gamma : \gamma < \lambda\} \quad \text{if } \lambda \text{ is a limit ordinal.}
\end{aligned}$$

Let us calculate a few examples.

$$\omega \cdot 2 = \omega(1 + 1) = \omega + \omega,$$

and similarly $\omega \cdot 3 = \omega + \omega + \omega$, $\omega \cdot 4 = \omega + \omega + \omega + \omega$, and so on. On the other hand,

$$2 \cdot \omega = \sup\{2n : n \in \mathbb{N}\} = \sup\{m : m \in \mathbb{N}\} = \omega.$$

(Think of $\alpha \cdot \beta$ as "α repeated β-many times.") Hence ordinal multiplication is not commutative either. Moreover, we have

$$\omega \cdot \omega = \sup\{\omega \cdot n : n \in \mathbb{N}\}.$$

Hence we can view $\omega \cdot \omega$ as the *limit* of the sequence

$$\omega, \ \omega + \omega, \ \omega + \omega + \omega, \ \omega + \omega + \omega + \omega, \ \ldots.$$

Similarly, we can now form the sequence

$$\omega, \ \omega \cdot \omega, \ \omega \cdot \omega \cdot \omega, \ \ldots.$$

What should the limit of this sequence be? If we let ourselves be guided by the analogy of the finite world of natural numbers, it ought to be

$$\omega^\omega.$$

Just as multiplication is obtained by iterating addition, exponentiation is obtained by iterating multiplication. We can do this for ordinals, too.

2.4. Ordinals, well-orderings, and axiom of choice

Exponentiation of ordinals.

$$\alpha^0 = 1,$$
$$\alpha^{\beta+1} = \alpha^\beta \cdot \alpha,$$
$$\alpha^\lambda = \sup\{\alpha^\gamma \colon \gamma < \lambda\} \quad \text{if } \lambda \text{ is a limit ordinal.}$$

By this definition, ω^ω is the limit of $\omega, \omega^2, \omega^3, \ldots$ indeed. Using exponentiation, we can form the sequence

$$\omega, \omega^\omega, \omega^{\omega^\omega}, \ldots.$$

The limit of this sequence is called ε_0. It is the least ordinal with the property that

$$\omega^{\varepsilon_0} = \varepsilon_0.$$

This property seems rather counterintuitive, since in the finite realm m^n is much larger than n as m and n grow larger and larger.

Ordinals, well-orderings and the axiom of choice. Let's look at the sequence of ordinals we have encountered so far:

$$0 < 1 < 2 < \cdots < \omega < \omega + 1 < \cdots < \omega + \omega < \cdots < \omega\omega < \cdots < \omega^\omega < \cdots < \varepsilon_0.$$

Property (O2) requires that the ordinal numbers be linearly ordered, and our initial list above clearly reflects this property. It turns out that the ordinals are a linear ordering of a special kind, a *well-ordering*.

Definition 2.14. Assume that $(S, <)$ is a linearly ordered set. We say that $(S, <)$ is a **well-ordering** if every non-empty subset of S has a $<$-least element.

In particular, this means that S itself must have a $<$-minimal element. Therefore, $\mathbb{Z}, \mathbb{Q}$, and $\mathbb{R}$ are not well-orderings. On the other hand, the natural numbers with their standard ordering are a well-ordering—in every non-empty subset of $\mathbb{N}$ there is a least number. If we restrict the rationals (or reals) to $[0, 1]$, we do not get a well-ordering, since the subset $\{1/n \colon n \geq 1\}$ does not have a minimal element *in the subset*.

The last example hints at an equivalent characterization of well-orderings.

Proposition 2.15. *A linear ordering* $(S, <)$ *is a well-ordering if and only if there does not exist an infinite descending sequence*

$$s_0 > s_1 > s_2 > \cdots$$

in S.

Proof. It is clear that if such a sequence exists, then the ordering cannot be a well-ordering, since the set $\{s_0, s_1, s_2, \ldots\}$ does not have a minimal element.

On the other hand, suppose $(S, <)$ is not a well-ordering. Then there exists a non-empty subset $M \subseteq S$ that has no minimal element with respect to $<$. We use this fact to construct a descending sequence $s_0 > s_1 > s_2 > \cdots$ in M.

Let s_0 be any element of M; then s_0 cannot be a minimum of M, since M does not have a minimum. Hence we can find an element $s_1 < s_0$ in M. But s_1 cannot be a minimum of M either, and hence we can find $s_2 < s_1$, and so on. $\square$

This characterization shows us that well-orderings have a strong *asymmetry*. While we can "count up" unboundedly through an infinite well-ordering, we cannot "count down" in the same way. No matter how we do it, after *finitely many steps* we reach the end, i.e. the minimal element.

We now verify that the ordinals are well-ordered by $<$.

Proposition 2.16. *Any set of ordinal numbers is well-ordered by* $<$.

Proof. Suppose S is a set of ordinals not well-ordered by $<$. Then there is an infinite descending sequence $\alpha_0 > \alpha_1 > \cdots$ in S. Let M be the set of all ordinals smaller than every element of the sequence, i.e.

$$M = \{\beta : \beta < \alpha_i \text{ for all } i\}.$$

By (O4), M has a least upper bound γ. Now γ has to be below all α_i as well, because otherwise every $\alpha_i < \gamma$ would be a smaller upper bound for M. Therefore, $\gamma \in M$.

By (O3), there exists a smallest ordinal greater than γ, namely its successor $\gamma + 1$. But $\gamma + 1$ cannot be in M, for otherwise γ would not be an upper bound for M. Hence there must exist an i such that

2.4. Ordinals, well-orderings, and axiom of choice

$\alpha_i < \gamma + 1$. But since $\gamma + 1$ is the smallest ordinal greater than γ, it follows that $\alpha_{i+1} < \alpha_i \leq \gamma$, contradicting that $\gamma \in M$. □

As we have mentioned above, the collection of all ordinals does not form a set. But if we fix any ordinal β, the **initial segment** of Ord up to β,

$$\text{Ord} \restriction_\beta = \{\alpha : \alpha \text{ is an ordinal and } \alpha < \beta\},$$

does form a set, and the proof as above shows that this initial segment is well-ordered.

Of course not every set that comes with a partial ordering is a well-ordering (or even a linear ordering). But if we are given the freedom to impose our own ordering, can every set be well-ordered?

This appears to be clear for finite sets. Suppose S is a non-empty finite set. We pick any element, declare it to be the minimal element, and pick another (different) element, which we declare to be the minimum of the remaining elements. We continue this process till we have covered the whole set. What we are really doing is constructing a bijection $\pi : \{0, 1, \ldots, n-1\} \to S$, where n is a natural number. The well-ordering of S is then given by

(2.1) $$s < t \ :\Leftrightarrow\ \pi^{-1}(s) < \pi^{-1}(t).$$

Can we implement a similar process for infinite sets? We have introduced the ordinals as a transfinite analogue of the natural numbers, so what we could try to do is find a bijection π between our set and an initial segment of the ordinals of the form

$$\{\alpha : \alpha \text{ is an ordinal and } \alpha < \beta\},$$

where β is an ordinal. Then, as can be easily verified, (2.1) again defines a well-ordering on our set.

Another way to think of a well-ordering of a set S is as an *enumeration of S indexed by ordinals*: If we let $s_\xi = \pi(\xi)$, then

$$S = \{s_0 < s_1 < s_2 < \cdots\} = \{s_\xi : \xi < \beta\}$$

for some ordinal β.

If a set is well-ordered, one can in turn show that the ordering is isomorphic to the well-ordering of an initial segment of Ord.

Proposition 2.17. *Suppose $(A, <)$ is a well-ordering. Then there exists a unique ordinal β and a bijection $\pi : A \to \{\alpha : \alpha < \beta\}$ such that for all $a, b \in A$*

$$a < b \iff \pi(a) < \pi(b).$$

We call β the **order type** of the well-ordering $(A, <)$.

You can try to prove Proposition 2.17 yourself as an exercise, but you will need to establish some more (albeit easy) properties of ordinals and well-orderings along the way. You can also look up a proof, for example in [**35**].

One can furthermore show that the order type of $\mathrm{Ord} \restriction_\beta$ is β. For this reason, ordinals are usually *identified with their initial segments in* Ord.

Returning to our question above, is it possible to well-order any set? The answer to this question is, maybe somewhat surprisingly, a hesitant "it depends".

The axiom of choice and the well-ordering principle. Intuitively, an argument for the possibility of well-ordering an arbitrary set S might go like this:

Let $S' = S$. If $S' \neq \varnothing$, let ξ be the least ordinal to which we have not yet assigned an element of S. Pick any $x \in S'$, map $\xi \mapsto x$, put $S' := S' \setminus \{x\}$, and iterate.

The problem here is the "pick any $x \in S'$". It seems an innocent step; after all, S' is assumed to be non-empty. But we have to look at the fact that we *repeatedly apply* this step. In fact, what we seem to assert here is the existence of a special kind of **choice function**: There exists a function f whose domain is the *set of all non-empty subsets of S*, $\mathcal{P}^0(S) = \{S' : \varnothing \neq S' \subseteq S\}$, such that for all $S' \in \mathcal{P}^0(S)$,

$$f(S') \in S'.$$

Indeed, equipped with such a function, we can formalize our argument above.

If $S = \varnothing$, we are done. So assume $S \neq \varnothing$. Put $s_0 = f(S)$.

2.4. Ordinals, well-orderings, and axiom of choice

Suppose now we have enumerated elements

$$\{s_\xi : \xi < \alpha\}$$

from S. If $S \smallsetminus \{s_\xi : \xi < \alpha\}$ is non-empty, put

$$s_\alpha = f(S \smallsetminus \{s_\xi : \xi < \alpha\}).$$

Now iterate. This procedure has to stop at some ordinal, i.e. there exists an ordinal β such that

$$S = \{s_\xi : \xi < \beta\}.$$

If not, that is, if the procedure traversed *all* ordinals, we would have constructed an injection $F : \mathrm{Ord} \to S$. Using some standard axioms about sets, this would imply that, since Ord is not a set, S cannot be a set (it would be as large as the ordinals, which form a proper class), which is a contradiction.

Does such a choice function exist? Most mathematicians are comfortable to assume this, or at least they do not feel that there is overwhelming evidence against it. It turns out, however, that the existence of general choice functions is a mathematical principle that cannot be reduced to or proved from other, more evident principles (this is the result of some seminal work on the foundations of mathematics first by Gödel [20] and then by Cohen [8, 9]). It is therefore usually stated as an *axiom*.

Axiom of choice (AC): Every family of non-empty sets has a choice function. That is, if $\mathcal{S}$ is a family of sets and $\varnothing \notin \mathcal{S}$, then there exists a function f on $\mathcal{S}$ such that $f(S) \in S$ for all $S \in \mathcal{S}$.

The axiom of choice is equivalent to the following principle.

Well-ordering principle (WO): Every set can be well-ordered.

We showed above that (AC) implies (WO). It is a nice exercise to show the converse.

Exercise 2.18. Derive (AC) from (WO).

There are some consequences of the axiom of choice that are, to say the least, puzzling. Arguably the most famous of these is the

Banach-Tarksi paradox: Assuming the axiom of choice, it is possible to partition a unit ball in $\mathbb{R}^3$ into finitely many pieces and rearrange the pieces so that we get *two* unit balls in $\mathbb{R}^3$.

Why has the Banach-Tarski paradox not led to an outright rejection of the axiom of choice, as this consequence clearly seems to run counter to our geometric intuition?

The reason is that our intuitions about notions such as *length* and *volume* are not so easy to formalize mathematically. The pieces obtained in the Banach-Tarksi decomposition of a ball are what is called *non-measurable*, meaning essentially that the concept of volume in Euclidean space as we usually think about it (the *Lebesgue measure*) is not applicable to these pieces.

For now, let us just put on record that the use of the axiom of choice may present some foundational issues. By using a choice function without specifying further the specific objects which are chosen, the axiom introduces a non-constructive aspect into proofs. For this reason, one often tries to clarify whether the axiom of choice is needed in its full strength, whether it can be replaced by weaker (and foundationally less critical) principles such as the axiom of countable choice (AC_ω) or the axiom of dependent choice (DC), or whether it can be avoided altogether (for example by giving an explicit, constructive proof).

The book by Jech [**36**] is an excellent source on many questions surrounding the axiom of choice.

2.5. Cardinality and cardinal numbers

We introduced ordinals as a continuation of the counting process through the transfinite. In the finite realm, one of the main purposes of counting is to establish cardinalities. We count a finite set by assigning its elements successive natural numbers. In other words,

2.5. Cardinality and cardinal numbers 65

to count a finite set S means to put the elements of S into a one-to-one correspondence with the set $\{0, \ldots, n-1\}$, for some natural number n. In this case we say S has cardinality n.

How can this be generalized to infinite sets? The basic idea is that:

Two sets have the same cardinality if there is a bijection (a mapping that is one-to-one and onto) between them.

For example, the sets $\{1, 2, 3, 4, 5\}$ and $\{6, 7, 8, 9, 10\}$ have the same cardinality. In the finite realm, it is impossible for a set to be a proper subset of another set yet have the same cardinality as the other set. This is no longer the case for infinite sets. The set of integers has the same cardinality as the set of even integers, as witnessed by the bijection $z \mapsto 2z$.

A very interesting case is $\mathbb{N}$ versus $\mathbb{N} \times \mathbb{N}$. While $\mathbb{N}$ is not a subset of $\mathbb{N} \times \mathbb{N}$, we can embed it via the mapping $n \mapsto (n, 0)$ as a proper subset of $\mathbb{N} \times \mathbb{N}$. But there is actually a bijection between the two sets, the **Cantor pairing function**

$$(x, y) \mapsto \langle x, y \rangle = \frac{(x+y)^2 + 3x + y}{2}.$$

Exercise 2.19. (a) Draw the points of $\mathbb{N} \times \mathbb{N}$ in a two-dimensional grid. Start at $(0, 0)$, which maps to 0, and find the point which maps to 1. Connect the two with an arrow. Next find the point that maps to 2, and connect it by an arrow to the point that maps to 1. Continue in this way. What pattern emerges?

(b) We can rewrite the pairing function as

$$\frac{(x+y)^2 + 3x + y}{2} = x + \frac{(x+y+1)(x+y)}{2}.$$

Recall that the sum of all numbers from 1 to n is given by

$$\frac{(n+1)n}{2}.$$

How does this help to explain the pattern in part (a)?

It can be quite hard to find a bijection between two sets of the same cardinality. The **Cantor-Schröder-Bernstein theorem** can be very helpful in this regard.

Theorem 2.20. *If there is an injection $f : X \to Y$ and an injection $g : Y \to X$, then X and Y have the same cardinality.*

You can find a proof in [35]. You can of course try proving it yourself, too.

Exercise 2.21. Use the Cantor-Schröder-Bernstein theorem to show that $\mathbb{R}$ and $[0,1]$ have the same cardinality.

Being able to map a set bijectively to another set is another important example of an equivalence relation (see Section 1.2). Let us write
$$A \sim B \colon \Leftrightarrow \text{ there exists a bijection } \pi : A \to B.$$

Exercise 2.22. Show that $\sim$ is an equivalence relation, that is, it is reflexive, symmetric, and transitive.

We could define the cardinality of a set to be its equivalence class with respect to $\sim$. (This would indeed be a proper *class*, not a set.) While this is mathematically sound, it makes thinking about and working with cardinalities rather cumbersome.

One way to overcome this is to pick a canonical representative for each equivalence class and then study the system of representatives.

In the case of cardinalities, what should be our representatives? For finite sets, we use natural numbers. For infinite sets, we can try to use ordinals, as they continue the counting process beyond the finite realm. Counting, in this generalized sense, means establishing a bijection between the set we are counting and an ordinal. If we assume the axiom of choice, the well-ordering principle ensures that every set can be well-ordered, so every set would have a representative. The only problem is that *an infinite set can be well-ordered in more than one way*.

Consider for instance the set of integers, $\mathbb{Z}$. We can well-order $\mathbb{Z}$ as follows:
$$0 < 1 < -1 < 2 < -2 < 3 < \cdots.$$
This gives a well-ordering of order type ω. But we could also proceed like this:
$$1 < -1 < 2 < -2 < 3 < -3 < \cdots < 0,$$

2.5. Cardinality and cardinal numbers

that is, we put 0 on top of all other numbers. This gives a well-ordering of order type $\omega+1$. Or we could put all the negative numbers on top of the positive integers:

$$0 < 1 < 2 < 3 < \cdots < -1 < -2 < -3 < \cdots,$$

which gives a well-ordering of type $\omega + \omega$.

This implies, in particular, that ω, $\omega + 1$, and $\omega + \omega$ all have the same cardinality. Recall that we identify ordinals with their initial segment, i.e. we put $\beta = \{\alpha \in \text{Ord}: \alpha < \beta\}$. Hence $\omega + 1 = \{0, 1, 2, \ldots, \omega\}$, and we can map $\omega + 1$ bijectively to ω as follows

$$\omega \mapsto 0, \quad 0 \mapsto 1, \quad 1 \mapsto 2, \quad \ldots$$

Exercise 2.23. Show that ω^ω has the same cardinality as ω.

To obtain a unique representative for each cardinality, we pick the *least ordinal* in each equivalence class. (Here it comes in very handy that the ordinals are well-ordered.)

Definition 2.24. An ordinal κ is a **cardinal** if for all ordinals $\beta < \kappa$, $\beta \nsim \kappa$.

For example, $\omega + 1$ is not a cardinal, while ω is—every ordinal below ω is finite and hence not of the same cardinality as ω. Thus, ω enjoys a special status in that it is the *first infinite cardinal*.

Exercise 2.25. Show that every cardinal greater than ω is a limit ordinal.

To define the **cardinality** of a set S, denoted by $|S|$, we now simply pick out the one ordinal among all possible order types of S that is a cardinal

$$|S| = \min\{\alpha: \text{there exists a well-ordering of } S \text{ of order type } \alpha\}$$
$$= \text{the unique cardinal } \kappa \text{ such that } S \sim \kappa.$$

Note that this definition uses the axiom of choice, since we have to ensure that each set has at least one well-ordering.

Exercise 2.26. Show that $|A| \le |B|$ if and only if there exists a one-to-one mapping $A \to B$.

How many cardinals are there? Infinitely many. This is Cantor's famous theorem.

Theorem 2.27. *For every set S, there exists a set of strictly larger cardinality.*

Proof. Consider $\mathcal{P}(S) = \{X : X \subseteq S\}$, the *power set* of S. The mapping $S \to \mathcal{P}(S)$ given by $s \mapsto \{s\}$ is clearly injective, so $|S| \leq |\mathcal{P}(S)|$.

We claim that there is no bijection $f : S \to \mathcal{P}(S)$. Suppose there were such a bijection f, that is, in particular,

$$\mathcal{P}(S) = \{f(x) : x \in S\}.$$

Every subset of S is the image of an element of S under f. To get a contradiction, we exhibit a set $X \subseteq S$ for which this is impossible, namely by letting

$$x \in X \iff x \notin f(x).$$

Now, if there were $x_0 \in S$ such that $f(x_0) = X$, then, by the definition of X,

$$x_0 \in X \iff x_0 \notin f(x_0) \iff x_0 \notin X,$$

a contradiction. This is a set-theoretic version of Cantor's *diagonal argument*. □

The power set operation always yields a set of higher cardinality. But by how much? Since the ordinals are well-ordered, so are the cardinals. We can therefore define, for any cardinal κ,

$$\kappa^+ = \text{the least cardinal greater than } \kappa.$$

Cardinal arithmetic. We now define the basic arithmetic operations on cardinals. Given cardinals κ and λ, let A and B be sets such that $|A| = \kappa, |B| = \lambda$, and $A \cap B = \varnothing$. Let

(2.2) $\quad \kappa + \lambda = |A \cup B|,$

(2.3) $\quad \kappa \cdot \lambda = |A \times B|,$

(2.4) $\quad \kappa^\lambda = |A^B| = |\{f : f \text{ maps } B \text{ to } A\}|.$

Exercise 2.28. Verify that the definitions above are independent of the choice of A and B.

2.5. Cardinality and cardinal numbers

The power operation 2^κ is particularly important because it coincides with the cardinality of the power set of κ:

$$2^\kappa = \text{cardinality of } \mathcal{P}(\kappa).$$

In some ways, cardinal arithmetic behaves just like the familiar arithmetic of real numbers.

Exercise 2.29. Let κ, λ, and μ be cardinals. Show that

$$(\kappa^\lambda)^\mu = \kappa^{\lambda \cdot \mu}.$$

But in other regards, cardinal arithmetic is very different.

Proposition 2.30. *Let κ and λ be infinite cardinals. Then*

$$\kappa + \lambda = \kappa \cdot \lambda = \max\{\kappa, \lambda\}.$$

Exercise 2.31. Prove Proposition 2.30.

Note that for many arguments involving cardinals and cardinal arithmetic, the axiom of choice is needed.

Alephs and the continuum hypothesis. Let us denote the cardinality of $\mathbb{N}$ by $\aleph_0$, i.e. any countable, infinite set has cardinality $\aleph_0$. Pronounced "aleph", $\aleph$ is the first letter of the Hebrew alphabet. The cardinal $\aleph_0$ is the smallest infinite cardinal. We know that the real numbers $\mathbb{R}$ are uncountable, i.e. $|\mathbb{R}| > \aleph_0$, and it is not hard to show (identifying reals with their binary expansions, which in turn can be interpreted as characteristic functions of subsets of $\mathbb{N}$) that $|\mathbb{R}| = 2^{\aleph_0}$. But is the cardinality of the reals actually the smallest uncountable cardinality? That is, is it true that

$$\aleph_0^+ = 2^{\aleph_0}?$$

This is the **continuum hypothesis (CH)**. Like the axiom of choice, the continuum hypothesis is independent over the most common axiom system for set theory, ZF. This means that the continuum hypothesis can be neither proved nor disproved in this axiom system. We will say more about independence in Chapter 4.

Since every cardinal has a successor cardinal (just like ordinals), we can use ordinals to index cardinals: We let

$$\aleph_1 = \aleph_0^+,$$

and more generally, for any ordinal α,
$$\aleph_{\alpha+1} = \aleph_\alpha^+.$$
We use ω_α instead of $\aleph_\alpha$ to denote the order type of the cardinal $\aleph_\alpha$. If λ is a limit ordinal, we define
$$\aleph_\lambda = \sup\{\omega_\alpha : \alpha < \lambda\}.$$

Exercise 2.32. Show that α_λ as defined above is indeed a cardinal. In other words, if S is a set of cardinals, so is the supremum of S.

Exercise 2.33. Show that every cardinal is an aleph, i.e. if κ is a cardinal, then there exists an ordinal α such that $\kappa = \aleph_\alpha$.

Generalized continuum hypothesis (GCH):
$$\text{For any } \alpha, \ \aleph_{\alpha+1} = 2^{\aleph_\alpha}.$$

If the GCH is true, it means that cardinalities are neatly aligned with the power set operation. The **beth function** $\beth_\alpha$ is defined inductively as
$$\beth_0 = \aleph_0, \quad \beth_{\alpha+1} = 2^{\beth_\alpha}, \quad \beth_\lambda = \sup\{\beth_\alpha : \alpha < \lambda\} \text{ for } \lambda \text{ a limit ordinal}.$$

That is, $\beth_\alpha$ enumerates the cardinalities obtained by iterating the power set operation, starting with $\mathbb{N}$. If the GCH holds, then $\beth_\alpha = \aleph_\alpha$ for all α.

2.6. Ramsey theorems for uncountable cardinals

Equipped with the notion of a cardinal, we can now attack the question of whether Ramsey's theorem holds for uncountable sets. It also makes sense now to consider colorings with *infinitely many colors*— the corresponding Ramsey statements are not trivially false anymore. We will also look at colorings of *sets of infinite tuples* over a set.

It is helpful to extend the arrow notation for these purposes. Recall that $N \longrightarrow (k)^p_r$ means that every r-coloring of $[N]^p$ has a monochromatic subset of size k. This is really a statement about cardinalities, which we can extend from natural numbers to cardinals.

Let κ, μ, η, and λ be cardinals, where $\mu, \eta \leq \kappa$.
$$\kappa \longrightarrow (\eta)^\mu_\lambda$$

2.6. Ramsey theorems for uncountable cardinals

means:

If $|X| \geq \kappa$ and $c: [X]^\mu \to \lambda$, then there exists $H \subseteq X$ with $|H| \geq \eta$ such that $c|_{[H]^\mu}$ is constant.

Here $[X]^\mu$ is the set of all subsets of X of cardinality μ:

$$[X]^\mu = \{D : D \subseteq X \text{ and } |D| = \mu\}.$$

The following lemma keeps track of the cardinality of $[X]^\mu$.

Lemma 2.34. *If $\kappa \geq \mu$ are infinite cardinals and $|A| = \kappa$, then*

$$[A]^\mu = \{D : D \subseteq A \text{ and } |D| = \mu\}$$

has cardinality κ^μ.

Proof. As $|A| = \kappa$, any element of κ^μ corresponds to a mapping $f : \mu \to A$, which is a subset of $\mu \times A$. Moreover, any such f satisfies $|f| = \mu$. Hence $\kappa^\mu \leq |[\mu \times A]^\mu| = |[A]^\mu|$, as $|\mu \times A| = |A|$.

On the other hand, we can define an injection $[A]^\mu \to A^\mu$: If $D \subseteq A$ with $|D| = \mu$, we can choose a function $f_D : \mu \to A$ whose range is D. Then the mapping $D \mapsto f_D$ is one-to-one. $\square$

As $\aleph_0$ is the smallest infinite cardinal, we can now write the infinite Ramsey theorem, Theorem 2.1, as

$$\aleph_0 \longrightarrow (\aleph_0)^p_r \quad \text{(for any } p, r \in \mathbb{N}\text{)}.$$

Finite colorings of uncountable sets. Does the infinite Ramsey theorem still hold if we pass to uncountable cardinalities?

Let us try to lift the proof from $\mathbb{N}$ to $\aleph_1$. To keep things simple, let us assume we are coloring pairs of real numbers with two colors. In the proof of $\aleph_0 \longrightarrow (\aleph_0)^2_2$, one proceeds by constructing a sequence of natural numbers

$$z_0, z_1, z_2, \ldots$$

along with a sequence of sets

$$\mathbb{N} = Z_0 \supseteq Z_1 \supseteq Z_2 \supseteq \cdots$$

such that

- $z_i \in Z_i$;
- for each i, the color of $\{z_i, z_j\}$ is the same for all $j > i$; and
- each Z_i is infinite.

It was possible to find these sequences because of the simple infinite pigeonhole principle: *If we partition an infinite set into finitely many parts, one of them must be infinite.*

This principle still holds for uncountable sets: Any finite partition of an uncountable set must have an uncountable part. In fact, we have something stronger:

In any partition of an uncountable set into countably many parts, one of the parts must be uncountable.

Using the language of cardinals, we can state and prove a formal version of this principle.

Proposition 2.35. *If κ is an uncountable cardinal and $f : \kappa \to \omega$, then there exists an $\alpha < \omega$ such that $|f^{-1}(\{\alpha\})| = \aleph_1$.*

Proof. Assume for a contradiction that $\kappa_\alpha = |f^{-1}(\{\alpha\})|$ is countable for all $\alpha < \omega$. Then
$$\kappa = \bigcup_{\alpha<\omega} \kappa_\alpha$$
would be a countable union of countable sets, which is countable—a contradiction. □

Looking at the countable infinite case, one might conjecture that an even stronger pigeonhole principle should be true, namely that there is an α such that $|f^{-1}(\{\alpha\})| = \kappa$. This is not quite so; it touches on an aspect of cardinals called *cofinality*. We will learn more about it when we look at large cardinals.

The pigeonhole principle is one instance where uncountable cardinals can behave rather differently from $\aleph_0$. We will see that this has consequences for Ramsey's theorem.

We return to Ramsey's theorem and try to prove $\aleph_1 \longrightarrow (\aleph_1)_2^2$.

We start with the usual setup. We choose $z_0 \in \aleph_1$ and look at all $z \in \aleph_1 \setminus \{z_0\}$ such that $\{z_0, z\}$ is red. If the set of such z is uncountable,

2.6. Ramsey theorems for uncountable cardinals

then we put $Z_1 = \{z : c(z_0, z) = \text{red}\}$. Otherwise, by the uncountable pigeonhole principle, $\{z : c(z_0, z) = \text{blue}\}$ is uncountable, and we let Z_1 be this set. We can now continue as usual inductively and construct the sequences $z_0, z_1, z_2, \ldots$ and $Z_1 \supseteq Z_2 \supseteq Z_3 \supseteq \cdots$, where each Z_i is uncountable. In the countable case, we were almost done, since it took only one more application of the (countable) pigeonhole principle to select an infinite homogeneous subsequence from the z_i. Now, however, we cannot do this, since we are looking for an *uncountable* homogeneous set. We therefore need to continue our sequence into the transfinite. How can this be done? We need to choose a "next" element of our sequence. We have learned in Section 2.4 that ordinals are made for exactly that purpose. Hence we would index the next element by z_ω. But what should the corresponding set Z_ω be? We required $Z_1 \supseteq Z_2 \supseteq Z_3 \supseteq \cdots$, and hence we should have that

$$Z_1 \supseteq Z_2 \supseteq Z_3 \supseteq \cdots \supseteq Z_\omega.$$

The only possible choice would therefore be

$$Z_\omega = \bigcap_i Z_i.$$

But this is a problem, because *the intersection of countably many uncountable nested sets is not necessarily uncountable*. Consider for instance the intersection of countably many open intervals

$$\bigcap_n (0, 1/n),$$

which is empty. Indeed, this obstruction is not a coincidence.

Proposition 2.36. *Ramsey's theorem does not hold for the real numbers:*

$$|\mathbb{R}| = 2^{\aleph_0} \not\to (2^{\aleph_0})^2_2.$$

We will, in fact, show something slightly stronger:

$$2^{\aleph_0} \not\to (\aleph_1)^2_2.$$

Of course, if the continuum hypothesis holds, this is equivalent to the previous statement.

Proof. The proof is based on the fact that a well-ordering of $\mathbb{R}$ must look *very different* from the usual ordering of the real line.

Using the axiom of choice, let $\prec$ be any well-ordering of $\mathbb{R}$. Hence we can write

$$\mathbb{R} = \{x_0 \prec x_1 \prec \cdots \prec x_\alpha \prec x_{\alpha+1} \prec \cdots\} \quad \text{with all } \alpha < 2^{\aleph_0}.$$

We define a coloring $c : [\mathbb{R}]^2 \to \{\text{red}, \text{blue}\}$. Let $y \neq z$ be real numbers and denote the usual ordering of $\mathbb{R}$ by $<_\mathbb{R}$. Let

$$c(y, z) = \begin{cases} \text{red} & \text{if } (y <_\mathbb{R} z \text{ and } y \prec z) \text{ or } (z <_\mathbb{R} y \text{ and } z \prec y), \\ \text{blue} & \text{otherwise.} \end{cases}$$

Here $<_\mathbb{R}$ denotes the usual order of $\mathbb{R}$. In other words, we color a set $\{y, z\}$ red if the two orderings $<_\mathbb{R}$ and $\prec$ agree for y and z. If they differ, we color the pair blue.

Assume for a contradiction that H is a homogeneous subset for c of size $\aleph_1$. We can write H as

$$H = \{x_{\alpha_0} \prec x_{\alpha_1} \prec \cdots \prec x_{\alpha_\xi} \prec \cdots\} \quad \text{with } \xi < \aleph_1.$$

If $c \upharpoonright_{[H]^2} \equiv \text{red}$, then we have by definition of c that also

$$x_{\alpha_0} <_\mathbb{R} x_{\alpha_1} <_\mathbb{R} \cdots <_\mathbb{R} x_{\alpha_\xi} <_\mathbb{R} \cdots,$$

that is, H gives us a $<_\mathbb{R}$-*increasing sequence of length* $\aleph_1$. If $c \upharpoonright_{[H]^2} \equiv$ blue, then we get a $<_\mathbb{R}$-decreasing sequence of length $\aleph_1$.

We claim that there cannot be such a sequence.

The rationals are dense in $\mathbb{R}$ with respect to $<_\mathbb{R}$, i.e. between any two real numbers is a rational number (not equal to either of them). If there were a strictly $<_\mathbb{R}$-increasing or strictly $<_\mathbb{R}$-decreasing sequence of length $\aleph_1$, there would also have to be a strictly $<_\mathbb{R}$-increasing/decreasing sequence of rational numbers of length $\aleph_1$, but this is impossible, since the rationals are countable. □

The essence of the proof lies in the fact that a homogeneous set would "line-up" the well-ordering $\prec$ with the standard ordering $<_\mathbb{R}$ of $\mathbb{R}$. If this line-up is too long (uncountable), we get a contradiction due to the fact that $\mathbb{R}$ contains $\mathbb{Q}$ as a dense "backbone" (under $<_\mathbb{R}$).

The proof also links back to the difficulties encountered earlier when trying to lift Ramsey's theorem to $2^{\aleph_0}$.

2.6. Ramsey theorems for uncountable cardinals   75

We are dealing with two orderings here: a well ordering $\prec$ of $\mathbb{R}$ and the familiar linear ordering $<_{\mathbb{R}}$ of the real line. Let us call the first one the "enumeration" ordering, since it determines the order in which we enumerate $\mathbb{R}$ and which element we choose next during our attempted construction—the $\prec$-least available. We do not know much about this ordering other than that it is a well-ordering. (In fact, there are some metamathematical issues that prevent us from proving that any explicitly defined function from $\mathbb{R}$ to an ordinal is a bijection.)

The standard ordering $<_{\mathbb{R}}$, on the other hand, is the "color" ordering, since going up or down along it determines whether we color red or blue.

Let us try to follow the construction of a homogeneous set in the proof of Theorem 2.1 and see where it fails for coloring c. Pick the $\prec$-first element of $Z_0 = \mathbb{R}$, say x_{α_0}. Next check whether the set $\{y\colon c(x_{\alpha_0}, y) = \text{red}\}$ is uncountable. This is the case, since there are uncountably many y "to the right" of x_{α_0}, and also uncountably many y not yet enumerated (these appear *after* x_{α_0} in the well-ordering). Hence we put $Z_1 = \{y\colon c(x_{\alpha_0}, y) = \text{red}\}$ and repeat the argument for Z_1: Pick the $\prec$-least element of Z_1 (which must exist since $\prec$ is a well-ordering) and observe that $\{y \in Z_1\colon c(x_{\alpha_0}, y) = \text{red}\}$ is again uncountable. Inductively, we construct an increasing sequence

$$x_{\alpha_0} <_{\mathbb{R}} x_{\alpha_1} <_{\mathbb{R}} x_{\alpha_2} <_{\mathbb{R}} \cdots$$

and a nested sequence of sets

$$Z_0 \supset Z_1 \supset Z_2 \supset \cdots$$

such that

$$(x_{\alpha_n}, \infty) \supseteq Z_{n+1}.$$

But if $x_{\alpha_n} \to \infty$ (which might well be the case), this implies

$$\bigcap_n Z_n = \emptyset,$$

and hence after ω-many steps we cannot continue our construction.

We could try to select the α_n a little more carefully; in particular, we could, for instance, let $x_{\alpha_{n+1}}$ be the $\prec$-least element of Z_{n+1} such that $x_{\alpha_n} <_{\mathbb{R}} x_{\alpha_{n+1}} < x_{\alpha_n} + 1/2^n$. This way we would guarantee that we could continue our construction beyond stage ω and into the transfinite. In fact, by choosing the x_α carefully enough, we can ensure that the construction goes on for β-many stages for *any fixed countable ordinal* β. But the cardinality argument of the proof above tells us it is impossible to do this for $\aleph_1$-many stages.

Exercise 2.37. Show that Proposition 2.36 generalizes to

$$2^\kappa \nrightarrow (\kappa^+)_2^2.$$

(*Hint:* Show that $\{0,1\}^\kappa$ has no increasing or decreasing sequence of length κ^+.)

An obvious question now arises: If we allow higher cardinalities κ beyond $2^{\aleph_0}$, does $\kappa \to (\aleph_1)_2^2$ become true eventually? The *Erdős-Rado theorem* shows that we in fact only have to pass to the next higher cardinality.

Theorem 2.38 (Erdős-Rado theorem).

$$(2^{\aleph_0})^+ \longrightarrow (\aleph_1)_2^2.$$

Compared to the counterexample in Proposition 2.36, the extra cardinal gives us some space to

> *set aside an uncountable set such that whenever we extend our current homogeneous set, we leave this set untouched, i.e. we do not add elements from it.*

In this way, we can now guarantee that the sets Z_α will have a non-empty, in fact uncountable, intersection.

This "setting aside" happens by virtue of the following lemma.

Lemma 2.39. *There exists a set* $R \subset (2^{\aleph_0})^+$ *of cardinality* $|R| = 2^{\aleph_0}$ *such that for every countable* $D \subseteq R$ *and for every* $x \in (2^{\aleph_0})^+ \setminus D$, *there exists an* $r \in R \setminus D$ *such that for all* $d \in D$,

(2.5) $$c(x,d) = c(r,d).$$

Informally, whenever we choose an x and a countable $D \subset R$, we can find a "replacement" for x in R that behaves in a color-identical manner with respect to D. This will enable us, in our construction

2.6. Ramsey theorems for uncountable cardinals

of a homogeneous set of size $\aleph_1$, to choose the x_α from a set of size $2^{\aleph_0}$, leaving a "reservoir" of uncountable cardinality.

Proof. We construct the set R by extending it step by step, adding the witnesses required by (2.5).

We start by putting $R_0 = 2^{\aleph_0}$. We have to ensure that (2.5) holds for *every* countable subset $D \subset R_0$ and every $x \in (2^{\aleph_0})^+ \smallsetminus D$. To simplify notation, let us put

$$c_x(y) = c(x, y).$$

Hence every x fixes a function $c_x : (2^{\aleph_0})^+ \smallsetminus \{x\} \to \{0, 1\}$. We are interested in the functions $c_x \upharpoonright_D$ for countable $D \subset R_0$. Each such function maps a countable subset of R_0 to $\{0, 1\}$.

We count the number of such functions. If we fix a countable $D \subset R_0$, there are at most $2^{\aleph_0}$-many ways to map D to $\{0,1\}$. By Lemma 2.34, there are

$$(2^{\aleph_0})^{\aleph_0} = 2^{\aleph_0 \cdot \aleph_0} = 2^{\max\{\aleph_0, \aleph_0\}} = 2^{\aleph_0}$$

countable subsets of $2^{\aleph_0}$. Therefore, there are at most

$$2^{\aleph_0} \cdot 2^{\aleph_0} = 2^{\aleph_0}$$

possible functions $c_x \upharpoonright_D$. (Note that while each $x \in (2^{\aleph_0})^+$ gives rise to such a function, many of them will actually be identical, by the pigeonhole principle.)

Therefore, we need to add at most $2^{\aleph_0}$-many witnesses to R_0, one $r \in (2^{\aleph_0})^+$ for each function $c_x \upharpoonright_D$ (of which there are at most $2^{\aleph_0}$-many). This gives us R_1, and R_1 in turn gives rise to new countable subsets D which we have to witness by possibly adding new elements from $(2^{\aleph_0})^+$ to R_1. But the crucial fact here is that the cardinality of R_1 is still $2^{\aleph_0}$, since $2^{\aleph_0} + 2^{\aleph_0} = 2^{\aleph_0}$, and therefore we can resort to the same argument as before, adding at most $2^{\aleph_0}$-many witnesses, resulting in a set R_2 of cardinality $2^{\aleph_0}$.

We have to run our construction into the transfinite. Let α be a countable ordinal, and assume that we have defined sets

$$R_0 \subseteq R_1 \subseteq \cdots \subseteq R_\beta \subseteq \cdots$$

for all $\beta < \alpha$, $|R_\beta| = 2^{\aleph_0}$. If α is a successor ordinal, $\alpha = \beta+1$, we define R_α by the argument given above, adding at most $2^{\aleph_0}$ new witnesses. If α is a limit ordinal, we put

$$R_\alpha = \bigcup_{\beta < \alpha} R_\beta.$$

This is a countable union of sets of cardinality $2^{\aleph_0}$, and hence is also of cardinality $2^{\aleph_0}$ (see Proposition 2.30).

Finally, put

$$R = \bigcup_{\alpha < \omega_1} R_\alpha.$$

We claim that this R has the desired property. First note that the cardinality of R is

$$\aleph_1 \cdot 2^{\aleph_0} = \max\{\aleph_1, 2^{\aleph_0}\} = 2^{\aleph_0}.$$

Let $D \subset R$ be countable and let $\dot{x} \in (2^{\aleph_0})^+ \setminus D$. The crucial observation is that

all elements of D must have been added by some stage $\alpha < \omega_1$, i.e. there exists an $\alpha < \omega_1$ such that $D \subseteq R_\alpha$.

If this were not the case, every stage would add a new element of D, which would mean D has at least ω_1-many elements, in contradiction to D being countable. But then the necessary witness for $c_x \restriction_D$ is present in $R_{\alpha+1}$; that is, there exists an $r \in R_{\alpha+1} \subseteq R$ such that $c_r \restriction_D = c_x \restriction_D$.

This completes the proof of the lemma. $\square$

We can now use the lemma to modify our construction of a homogeneous set H.

Proof of the Erdős-Rado theorem. Let x^* be an arbitrary element of $(2^{\aleph_0})^+ \setminus R$. This will be our "anchor point". Choose $x_0 \in R$ arbitrary.

Suppose now, given $\alpha < \omega_1$, we have chosen x_β for all $\beta < \alpha$.
Let

$$D = \{x_\beta : \beta < \alpha\}.$$

This is a countable set (since $\alpha < \omega_1$). By Lemma 2.39, there exists an $r \in R$ such that $c_{x^*} \restriction_D = c_r \restriction_D$. Put $x_\alpha = r$.

2.6. Ramsey theorems for uncountable cardinals 79

By the pigeonhole principle, there exists $i \in \{0, 1\}$ such that

$$H = \{x_\alpha \colon \alpha < \omega_1,\, c(x^*, x_\alpha) = i\}$$

is uncountable. We claim that $c\restriction_{[H]^2} \equiv i$, i.e. H is homogeneous for c. For suppose $x_\zeta, x_\xi \in H$ and $\zeta < \xi$. Then, by definition of x_ξ,

$$c(x_\xi, x_\zeta) = c_\xi(x_\zeta) = c_{x^*}(x_\zeta) = c(x^*, x_\alpha) = i.$$

□

Note that the proof becomes quite elegant once we have proved the lemma. After we ensured the existence of the set R, we can work in some sense "backwards": We choose a single "anchor point" x^* from the reserved set $(2^{\aleph_0})^+ \smallsetminus R$. You can think of x^* as always being the next point chosen in the sense of the standard construction, only then to be replaced by an element from R which behaves exactly like it in terms of color pairings with the already constructed x_β.

Note also that we do not need to construct a sequence of shrinking sets Z_α anymore. In the standard construction, the Z_α represent the reservoir from which the next potential elements of the homogeneous set are chosen. They are no longer needed since x^* is always available (as explained above).

We also do not need to keep track of the color choices we made along the way, as x^* does this job for us, too. For example, if $c(x^*, x_0) = 0$, it follows by construction that $c(x_\beta, x_0) = 0$ for all $\beta > 0$, which in the previous constructions means that we restrict the Z to all elements which color 0 with x_0.

Exercise 2.40. Generalize the proof of the Erdős-Rado theorem (and Lemma 2.39) to show that

$$\beth_n^+ \longrightarrow (\aleph_1)_{\aleph_0}^{n+1}.$$

Infinite colorings. The Erdős-Rado theorem holds for countable colorings, too (see Exercise 2.40). What else can we say about infinite colorings? Clearly, the number of colors should be smaller than the set we are trying to color. For example,

$$\aleph_0 \nrightarrow (2)_{\aleph_0}^1.$$

But even if we make the colored set larger than the number of colors, this does not mean we can find even a finite homogeneous set.

Proposition 2.41. *For any infinite cardinal κ,*
$$2^\kappa \not\to (3)^2_\kappa.$$

Proof. We define $c : [2^\kappa]^2 \to \kappa$ as follows. An element of 2^κ corresponds to a $\{0,1\}$-sequence $(x_\beta : \beta < \kappa)$ of length κ (recall that 2^κ is the cardinality of the power set of κ, and every element of the power set can be coded by its characteristic sequence; $(x_\beta : \beta < \kappa)$ is such a characteristic sequence). Given two such sequences $(x_\beta) \neq (y_\beta)$, we let
$$c((x_\beta), (y_\beta)) = \text{the least } \alpha < \kappa \text{ such that } x_\alpha \neq y_\alpha.$$
Now assume that $(x_\beta), (y_\beta), (z_\beta)$ are pairwise distinct. Let $c((x_\beta), (y_\beta)) = \alpha$. Without loss of generality, $x_\alpha = 0$ and $y_\alpha = 1$. Now $z_\alpha \in \{0,1\}$, so either $z_\alpha = x_\alpha$ or $z_\alpha = y_\alpha$. In the first case $c((x_\beta), (z_\beta)) \neq \alpha$ and in the second case $c((y_\beta), (z_\beta)) \neq \alpha$. Therefore, there cannot exist a c-homogeneous subset of size 3. □

2.7. Large cardinals and Ramsey cardinals

The results of the previous section make the set of natural numbers stand out among the infinite sets not only because $\aleph_0$ is the first infinite cardinal but also for another reason: With respect to finite colorings of finite tuples, the natural numbers *admit a homogeneous subset of the same size*, or, in the Ramsey arrow notation,
$$\aleph_0 \longrightarrow (\aleph_0)^p_r$$
for any positive integers p and r.

In the previous section, we saw that this is no longer true for $\aleph_1$ (Proposition 2.36). In fact, for *any* infinite cardinal κ,
$$2^\kappa \not\to (\kappa^+)^2_2,$$
so in particular
$$2^\kappa \not\to (2^\kappa)^2_2$$
for any infinite cardinal κ.

This in turn means that *any* cardinal λ that can be written as $\lambda = 2^\kappa$ cannot satisfy $\lambda \longrightarrow (\lambda)^2_2$. But are there any cardinals that *cannot*

2.7. Large cardinals and Ramsey cardinals

be written this way? $\aleph_0$, the cardinality of $\mathbb{N}$, is such a cardinal—the power set of a finite set is still finite. But other than $\mathbb{N}$?

Definition 2.42. A cardinal λ is a **limit cardinal** if $\lambda = \aleph_\gamma$ for some limit ordinal γ. λ is a **strong limit cardinal** if for all cardinals $\kappa < \lambda$, $2^\kappa < \lambda$.

Any strong limit cardinal is a limit cardinal. For if λ is a successor cardinal, then $\lambda = \aleph_{\alpha+1} = \aleph_\alpha^+ \leq 2^{\aleph_\alpha}$. Is being *strong limit* truly stronger than being limit? Well, it depends. If the GCH holds, then $2^{\aleph_\alpha} = \aleph_\alpha^+ = \aleph_{\alpha+1}$ for all α, and therefore every limit cardinal is actually a strong limit cardinal.

For now just let us assume that κ *is* a strong limit cardinal. Is this sufficient for
$$\kappa \longrightarrow (\kappa)_2^2?$$
Take for example $\aleph_\omega$. This is clearly a limit cardinal and, if the GCH holds, also a strong limit cardinal. Does it hold that
$$\aleph_\omega \longrightarrow (\aleph_\omega)_2^2?$$
The problem is that we can "reach" $\aleph_\omega$ rather "quickly" from below, since
$$\aleph_\omega = \bigcup_{n<\omega} \aleph_n.$$
We can use this fact to devise a coloring of $\aleph_\omega$ that cannot have a homogeneous subset of size $\aleph_\omega$. Namely, let us put, for each $n < \omega$,
$$X_{n+1} = \aleph_{n+1} \smallsetminus \aleph_n.$$
Then $\aleph_\omega$ is the disjoint union of the X_n, and the cardinality of each X_n is strictly less than $\aleph_\omega$. Now define a coloring $c : [\aleph_\omega]^2 \to \{0,1\}$ by
$$c(x,y) = \begin{cases} 1 & \text{if } x \text{ and } y \text{ are in different } X_n, \\ 0 & \text{if } x \text{ and } y \text{ are in the same } X_n. \end{cases}$$
Let $H \subset \aleph_\omega$ be a homogeneous subset for c. If $c\restriction_{[H]^2} \equiv 1$, then no two elements of H can be in the same X_n, but there are only $\aleph_0$-many X_n, and hence H is countable. If $c\restriction_{[H]^2} \equiv 0$, then all elements of H have to be in the same X_n, but as noted above, $|X_n| < \aleph_\omega$ for each n.

The proof works in general for any cardinal κ that we can reach in fewer than κ steps. This brings us to the notion of *cofinality*.

Definition 2.43. The **cofinality** of a limit ordinal α, $\mathrm{cf}(\alpha)$, is the least ordinal β such that there exists an increasing sequence $(\alpha_\gamma)_{\gamma<\beta}$ of length β such that $\alpha_\gamma < \alpha$ for all α and

$$\alpha = \lim_{\gamma \to \beta} \alpha_\gamma = \sup\{\xi_\gamma : \gamma < \beta\}.$$

Obviously, we always have $\mathrm{cf}(\alpha) \leq \alpha$. Here are some examples as an exercise.

Exercise 2.44. Prove the following cofinalities:

(i) $\mathrm{cf}(\omega) = \omega$,

(ii) $\mathrm{cf}(\omega + \omega) = \omega$,

(iii) $\mathrm{cf}(\alpha) = \omega$ for every countable, infinite α,

(iv) $\mathrm{cf}(\omega_1) = \omega_1$ (if we assume AC),

(v) $\mathrm{cf}(\omega_\omega) = \omega$.

The last statement can be generalized to

$$\mathrm{cf}(\omega_\lambda) = \mathrm{cf}(\lambda),$$

where λ is any limit ordinal.

If κ is an infinite cardinal and $\mathrm{cf}(\kappa) < \kappa$, it means that κ can be reached from below by means of a "ladder" that has fewer steps than κ. Such a cardinal is called **singular**. If $\mathrm{cf}(\kappa) = \kappa$, κ is called **regular**. Hence $\aleph_0$ and $\aleph_1$ are regular cardinals, while $\aleph_\omega$ is singular. Assuming the axiom of choice, one can show that every successor cardinal, that is, a cardinal of the form $\aleph_{\alpha+1}$, is a regular cardinal.

It seems much harder for a limit cardinal to be regular. For this to be true, the following must hold:

$$\aleph_\lambda = \mathrm{cf}(\aleph_\lambda) = \mathrm{cf}(\lambda) \leq \lambda.$$

But since clearly $\aleph_\lambda \geq \lambda$, this means that

if $\aleph_\lambda$ is regular (λ limit), then $\aleph_\lambda = \lambda$.

This seems rather strange. Going, for example, from $\aleph_0$ to $\aleph_1$, we traverse ω_1-many ordinals, but the jump "costs" only one step in terms of cardinals. That means we have to go a *long, long way* if we ever want to "catch up" with the alephs again.

2.7. Large cardinals and Ramsey cardinals

Anyway, let us capture the notions of large cardinals we have found so far in a formal definition.

Definition 2.45. Let κ be an uncountable cardinal.

(i) κ is **weakly inaccessible** if it is regular and a limit cardinal.

(ii) κ is **(strongly) inaccessible** if it is regular and a strong limit cardinal.

(iii) κ is **Ramsey** if $\kappa \longrightarrow (\kappa)_2^2$.

Usually, strongly inaccessible cardinals are called simply *inaccessible*. Our investigation leading up to Definition 2.45 now yields the following.

Theorem 2.46. *Every Ramsey cardinal is inaccessible.*

Do Ramsey cardinals exist? We cannot answer this question here, nor will we in this book. In fact, in a certain sense we cannot answer this question *at all*. More precisely, the *existence of Ramsey cardinals cannot be proved in* ZF. As mentioned before in Section 2.5, ZF is a common axiomatic framework for set theory in which most of contemporary mathematics can be formalized. We will say more about axiom systems and formal proofs in Chapter 4.

Chapter 3

Growth of Ramsey functions

3.1. Van der Waerden's theorem

In this chapter, we return from the infinite realm to the finite, but we will see how we can *approximate* the infinite through *really big* finite numbers. This may sound absurd, but will hopefully make some sense by the end of Chapter 4.

We have seen in Chapter 1 that the diagonal Ramsey numbers $R(n) = R(n,n)$ grow rather fast. Nobody knows how fast exactly, but the upper and lower bounds proved in Chapter 1 tell us that the growth is exponential of some sort. Loosely speaking, there has to be a lot of chaos to guarantee the existence of a little bit of order inside.

It turned out that many other results in Ramsey theory exhibit a similar behavior. Arguably the most famous theorem of this kind (which actually preceded Ramsey's theorem by a few years) is van der Waerden's theorem on arithmetic progressions [68]. We will see, in fact, that the analysis of the associated *van der Waerden numbers* gives rise to an interesting metamathematical perspective.

An **arithmetic progression** (AP) of length k is a sequence of the form

$$a, \quad a+m, \quad a+2m, \quad \ldots, \quad a+(k-1)m,$$

where a and m are integers. We will call arithmetic progressions of length k simply *k-APs*.

Theorem 3.1 (Van der Waerden's theorem)**.** *Given integers $r, k \geq 1$, for any r-coloring of $\mathbb{N}$ there exists a monochromatic k-AP.*

In other words, in any finite coloring of the natural numbers we can find arithmetic progressions of arbitrary length. It is important to be aware that "arbitrary length" here means "arbitrary *finite* length". It is not true that any finite coloring of $\mathbb{N}$ has a monochromatic arithmetic progression of infinite length.

Exercise 3.2. Construct a 2-coloring of $\mathbb{N}$ such that neither color contains an infinite arithmetic progression.

(*Hint:* Alternate between longer and longer blocks of 1's and 2's.)

The exact location of the first k-AP will of course depend on the particular coloring. It is, however, possible to bound this first occurrence from above *a priori*, that is, independent of any coloring.

Theorem 3.3 (Van der Waerden's theorem, finite version)**.** *Given integers $r, k \geq 1$, there exists a number W such that any r-coloring of $\{1, \ldots, W\}$ has a monochromatic k-AP.*

The smallest possible W is called the **van der Waerden number** for k and r, and in the following we will denote it by $W(k, r)$.

That the finite version is a consequence of the first version is not completely obvious. It could be that for every n, we could find *one specific coloring* of $\{1, \ldots, n\}$ such that no monochromatic k-AP exists for that coloring. However, we could, as in the proof of the finite Ramsey theorem via compactness, collect these examples in a tree and find an infinite path, yielding a coloring of $\mathbb{N}$. A similar argument to that in the Ramsey case would then yield a contradiction.

Exercise 3.4. Give a detailed derivation of the finite version of van der Waerden's theorem from Theorem 3.1 using compactness. Follow the template outlined at the end of Section 2.2.

3.1. Van der Waerden's theorem

Van der Waerden himself gave a wonderful account of how he (together with E. Artin and O. Schreier) found the proof of the theorem that now bears his name [69, 70]. We will try to follow his exposition here.

"Start with the very simple examples" (Hilbert).

One easily sees that $W(2, r) = r + 1$. Any two numbers x and y form an arithmetic progression of length 2, by setting $a = x$ and $m = y - x$. Thus, to have a monochromatic 2-AP it suffices to have two numbers of the same color. But by the pigeonhole principle, any r-coloring of $\{1, \ldots, r+1\}$ has at least two numbers of the same color. And we also see that the bound $r + 1$ is optimal in this case.

As the case of $r = 2$ and $k = 3$ is not nearly as simple, van der Waerden first checked some cases by hand (literally using pencil and paper, we have to guess, as the year was 1927 and no computers were available). He found a 2-coloring of $\{1, \ldots, 8\}$ that does not have a monochromatic 3-AP (Figure 3.1):

Figure 3.1. A coloring of $\{1, \ldots, 8\}$ with no monochromatic 3-AP[1]

But then he saw that no matter how you color $\{1, \ldots, 9\}$ ($2^9 = 512$ possible colorings), there will always be a monochromatic 3-AP. In other words, $W(3, 2) = 9$. Keep in mind, though, that van der Waerden was originally trying to prove Theorem 3.1, not the finite version, Theorem 3.3. As we outlined above, the mere existence of $W(3, 2)$ is not clear at all from the statement of Theorem 3.1. But the three mathematicians were able to deduce Theorem 3.3 given that Theorem 3.1 holds by an argument similar to the compactness argument we have used. (They did not call it *compactness* though.)

---

[1] Our diagrams follow the design of van der Waerden [69]. In the case of two colors, the top line will always represent the color blue and the bottom one always red.

Two colors versus many.

Artin, Schreier, and van der Waerden thenceforth tried to prove the existence of the numbers $W(k,r)$ through induction. Artin remarked that the finite version of the theorem made an induction approach easier, since one could use the existence of $W(k,r)$, which is a specific natural number, in an attempt to establish the existence of $W(k+1,r)$.

Artin made another key observation:

(3.1) *If $W(k,2)$ exists for every k, then $W(k,r)$ exists for all r and k.*

In other words, if we know that the finite theorem holds for $r=2$ and *all* k, then it holds for *all* r and k.

Take for example $r=3$ and consider a 3-coloring of $\mathbb{N}$. This induces a 2-coloring if we identify colors 1 and 2. For this 2-coloring we know there exists a monochromatic arithmetic progression of length $W(k,2)$. If the color of this arithmetic progression is 3, then we are done (since $W(k,2) \geq k$). Otherwise we have a 2-colored arithmetic progression

$$a, \quad a+m, \quad a+2m, \quad \ldots, \quad a+(W(k,2)-1)m,$$

not quite what we want yet. But if we renumber the terms of our progression as $1, 2, 3, \ldots$, we induce a 2-coloring of $\{1, \ldots, W(k,2)\}$. By definition of $W(k,2)$ (and the assumption that it exists), there exists a monochromatic k-AP for this coloring. This k-AP in turn translates into a *monochromatic sub-k-AP* of the first progression. We therefore have shown that $W(k,3) \leq W(W(k,2),2)$. It should now be clear how this proof can be continued by induction.

Note, however, that in order to show that $W(k,3)$ exists, we have used the existence of $W(l,2)$ for potentially much larger l. We have to keep this in mind when we try to prove the existence of $W(k,r)$ by some form of double induction, as we have to avoid circular arguments (e.g., using the existence of $W(l,2)$ to prove the existence of $W(k,3)$ for $k < l$, but using $W(k,3)$ in turn to establish the existence of $W(l,2)$).

3.1. Van der Waerden's theorem

Block progressions.

Being able to use more than two colors turned out to play a crucial role in establishing the existence of $W(k, 2)$ by induction. The colors occur as "block colors", in the following sense. Suppose we have a 2-coloring of $\mathbb{N}$. For some fixed $m \geq 1$, consider now consecutive blocks of m numbers, i.e.

$$1, 2, \ldots, m, \quad m+1, m+2, \ldots, 2m, \quad 2m+1, m+2, \ldots, 3m, \quad \ldots$$

Every one of those blocks has a corresponding color sequence of length m. For example, if $m = 5$, then

$$\begin{array}{c} \text{─┼─────┼─┼─} \\ \text{──┼─┼─────} \end{array} \quad \text{or, in short,} \quad 1\,2\,2\,1\,1$$

would be a possible color block. There are 2^m such potential color blocks overall. The key idea is to

(3.2)   *regard an m-block as a **single entity** and consider these entities 2^m-colored*

Artin suggested that one could apply an induction hypothesis for $W(k-1, 2^m)$, that is, the existence of $(k-1)$-many m-blocks in arithmetic progression, each of which has the same m-color pattern. It was, however, unclear at that point how exactly this might be helpful.

Van der Waerden's breakthrough.

Being able to use more than two colors to establish the existence of $W(k, 2)$ proved crucial for the proof, together with the following principle, which we applied above:

(3.3)   *Any arithmetic progression "inside" an arithmetic progression is again an arithmetic progression.*

Let us try to deduce that $W(3, 2)$ exists, by a formal deduction instead of a brute force argument where we simply check all the cases.

It is helpful to imagine an evil adversary who counters every guess we make by choosing a coloring that makes it as hard as possible for us to find monochromatic arithmetic progressions. We have to corner him with our logic so that eventually he has no choice but to reveal a monochromatic 3-AP.

Inspecting the color of the numbers $1, 2, 3$, our opponent has to reveal a 2-AP, by the pigeonhole principle (recall that any two numbers with the same color form a monochromatic 2-AP). Without loss of generality, let us assume we are presented with the following coloring:

$$\begin{array}{ccc} 1 & 2 & 3 \end{array}$$

In other words, we have a blue 2-AP $\{1, 3\}$. The easiest way to a monochromatic 3-AP from there would be a blue 5:

But of course our opponent will not give us this easy victory and thus color 5 red:

$$\begin{array}{ccccc} 1 & 2 & 3 & 4 & 5 \end{array}$$

In fact, as you can easily check, there are plenty of ways to 2-color $\{1, \ldots, 5\}$ without having a monochromatic 3-AP. So let us assume that no 5-block of the form $\{5m + 1, 5m + 2, \ldots, 5(m + 1)\}$ contains a monochromatic 3-AP. But, as we will now see, we can turn our opponent's strategy—denying us the easy victory inside a 5-block—against him using a nice trick.

This is where the *block-coloring* idea comes in: A 2-colored 5-block represents one of $2^5 = 32$ possible patterns. Think of each pattern as a "color" that is a special blend of red and blue, determined by how often each color occurs and where it occurs within the five positions. The pigeonhole principle now guarantees that

3.1. Van der Waerden's theorem

among the first 33 5-blocks (i.e. the numbers $\{1, \ldots, 165\}$ divided into consecutive blocks of 5 numbers), one *color* (i.e. 5-pattern) must occur twice.

To make the argument more concrete, suppose that the blocks $B_8 = \{36, 37, 38, 39, 40\}$ and $B_{21} = \{101, 102, 103, 104, 105\}$ have the same 2-coloring pattern. (There is nothing particular about these two blocks; the argument works just as well for any other pair, with slightly adjusted numbers.) Now, as above, if we consider only the color of B_8, the pigeonhole principle tells us that two of 36, 37, and 38 must have the same color. Denote these elements by i and j so that $i \in \{36, 37\}$ and $j \in \{37, 38\}$. The step-width of this 2-AP is $m = j - i$. We also have $i + 2m \leq 40$, so we could theoretically complete this into a 3-AP inside B_8, but as we said, our opponent will not grant us this easy victory.

Let us assume, again for illustrative purposes, that $i = 36$ and $j = 38$ are colored blue, while $i + 2m = 40$ is colored red. B_{21} has the same coloring pattern and, depending on how 39 is colored, we might have the following picture:

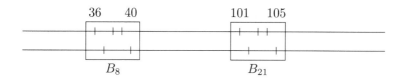

Figure 3.2. The 5-blocks B_8 and B_{21} have the same coloring patterns.

The two blocks together still do not give us a 3-AP. But the picture changes when we consider the *3-AP of 5-blocks generated by* B_8, B_{21}, *and* B_{34}. In other words, we consider a 3-AP of blocks corresponding to their indices.

Of course, the coloring pattern of B_{34} might be completely different from that of B_8 and B_{21}. But the crucial fact, as simple as it may sound, is that the last element of B_{34}, 170, is assigned a color. How would this help us? Let's look at the three blocks together:

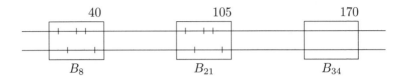

Figure 3.3. Extending the block-2-AP B_8, B_{21} to a block-3-AP B_8, B_{21}, B_{34}

Since the blocks are in arithmetic progression, the numbers inside the blocks are approximately in arithmetic progression. If we take the kth number from each block ($1 \le k \le 5$), we obtain in fact an exact arithmetic progression, such as $37, 102, 166$ (the second element of each block). This is the principle stated in (3.3): Any arithmetic progression inside an arithmetic progression is an arithmetic progression. But there are other ones: the progression $36, 102, 168$, for example, consisting of the first element of B_8, the second of B_{21}, and the third of B_{34}, or $36, 103, 170$ (the first, third, and fifth elements, respectively). The one that matters for us is the one corresponding to the monochromatic 2-AP in B_8 and B_{21}. In our example, this was the one given by the first and third elements (36 and 38, or 101 and 103, respectively). Now we "stretch out" this 3-AP over all three blocks: $36, 103, 170$.

Is this a monochromatic 3-AP? Not necessarily, as 170 might be colored red. But in this case $40, 105, 170$ is a red 3-AP.

By making 170 the "focal point" of two 3-APs, both of which are monochromatic in their first two terms (and with different colors), we have cornered our opponent. No matter what color he chooses for 170, it will complete a monochromatic 3-AP.

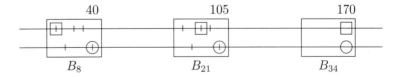

Figure 3.4. No matter how 170 is colored, we will get a monochromatic 3-AP: either $36, 103, 170$ (blue) or $40, 105, 170$ (red).

3.1. Van der Waerden's theorem

It should be clear how to reproduce this argument for other instances. In particular, we see that in the worst case, we need to consider the block-3-AP B_1, B_{33}, B_{65}. (This is the one that is most spread out.) In other words, our argument shows that

$$W(3,2) \leq 65 \cdot 5 = 325.$$

While this seems a rather large bound considering one can establish $W(3,2) = 9$ by checking all cases, the way we arrived at it can be generalized to higher orders.

Let us try to capture the essential steps in this argument.

1. *Choose a block size.* Our block size was 5 since it guaranteed the existence of a monochromatic 2-AP that can be extended to a 3-AP (not necessarily monochromatic) *within* the block.

2. *Find an arithmetic progression of block-coloring patterns.* There are $2^5 = 32$ possible 2-colorings of 5-blocks. Hence among 33 5-blocks, two must be the same. These blocks form a 2-AP of blocks.

3. *Extend the arithmetic progression of block patterns by one block.* The additional block may not have the same coloring pattern, but we will use it to "force" a monochromatic AP in the next step.

4. *Consider arithmetic progressions of numbers inside the block progression.* One element of the additional block will be the "focal point" of a monochromatic AP, either by collecting the elements in each block at a constant position, or by extending a "diagonal" AP, such as, in our example, elements 1, 3, and 5 from blocks B_8, B_{21}, and B_{34}, respectively.

Exercise 3.5. Follow the template above to argue that $W(4,2)$ exists. You may assume that all numbers of the form $W(3,r)$ exist. It helps to make a drawing like we did above.

1. We need a block size that guarantees the existence of a monochromatic 3-AP that can be extended to a 4-AP (not necessarily monochromatic). Express this in terms of $W(3,2)$.

2. Let M be the block size. There are 2^M ways to color such a block. But now we need to find not only two but *three* blocks

with an identical color pattern (since we want to extend to a 4-AP). Above we used the pigeonhole principle, but we can also say we used the existence of $W(2,32) = 33$. Which van der Waerden number $W(3,\cdot)$ would we use in this case?

3. How far would we have to go next to extend this 3-AP of blocks by one block?

4. Find the focal point of two monochromatic 3-APs and argue as above that the focal point must extend one of them to a monochromatic 4-AP.

Working through this example, it hopefully becomes clear how one can apply the template to show that $W(k,2)$ exists for an arbitrary k, assuming that $W(k-1,r)$ exists for all r.

Can this method be adapted to show the existence of $W(k,r)$ for $r > 2$? Before we think about this, did we not show in (3.1) that once we have proved the result for $r = 2$, it follows for all $r > 2$? Yes, but the crux is that in the proof of (3.1), we needed the existence of $W(W(k,2),2)$ to establish the existence of $W(k,3)$. And above we saw that to establish the existence of $W(4,2)$, we need the existence of numbers $W(3,r)$. Hence we cannot combine the two proofs, as it would yield a circular argument. Observation (3.1) only assured Artin, Schreier, and van der Waerden that an inductive approach using the block-coloring method was *possible*, since there would be no "holes" in the table of numbers $W(k,r)$. But the actual proof of going from $W(k,2)$ to $W(k,3)$ would have to proceed differently to avoid the circularity outlined above.

We again follow van der Waerden in demonstrating that $W(3,3)$ exists. We start by choosing a block size that ensures the existence of a monochromatic 2-AP with a possible extension to a 3-AP. As we have three colors now, a monochromatic 2-AP will appear among four numbers instead of three, and to extend it to a 3-AP we need a block size of 7 instead of 5.

We can apply the previous line of reasoning and get a 3-AP of blocks, the first two of which have an identical coloring pattern. These three blocks span, in the worst case, the numbers from 1 to $7 \cdot 3^7 + 7 + 7 \cdot 3^7 = 7 \cdot (2 \cdot 3^7 + 1)$. But the "focal point" argument does not work

3.1. Van der Waerden's theorem

anymore (or at least not right away), since the "focal number" could now be colored with the third color, say green.

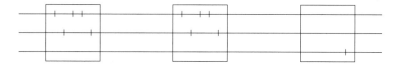

Figure 3.5. Escaping a monochromatic 3-AP by coloring the focal point green

But we can apply the block argument again—to the big block of size $M = 7 \cdot (2 \cdot 3^7 + 1)$. There are

$$3^M = 3^{7 \cdot (2 \cdot 3^7 + 1)} \text{ possible 2-colorings of such a block.}$$

Hence, among $3^M + 1$ such blocks, we must see two with the same coloring pattern.

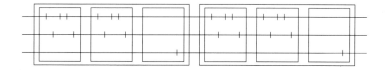

Figure 3.6. A 2-AP of blocks each containing a 3-AP of blocks

And again, they define a 2-AP of blocks. Once more, we extend this 2-AP to a 3-AP:

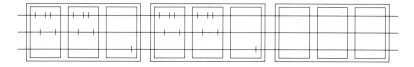

Figure 3.7. Extending the 2-AP of "big blocks" to a 3-AP

We do not know what the coloring of the third big block is, but arguing as before, one element in the third (inner) 7-block of the third (outer) M-block becomes the focal point of three arithmetic progressions, as indicated in the following picture:

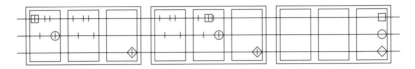

Figure 3.8. Finding a focal point of three different 3-APs

Each of the three arithmetic progressions is monochromatic in the first two terms, and each has a different color. Hence no matter what color the focal point is, it will complete a monochromatic 3-AP.

Exercise 3.6. Show that the above argument establishes that $W(3,3) \leq 5 \cdot 10^{14616}$.

(*Hint:* $5 \cdot 10^{14616} > (2 \cdot 3^{7 \cdot (2 \cdot 3^7 + 1)} + 1) \cdot (2 \cdot 3^7 + 1) \cdot 7$.)

Note that our argument uses only the pigeonhole principle (i.e. the existence of $W(2,r)$ for $r = 3^M$). The existence of $W(k,r)$ for $k > 2$ and $r \geq 2$ is not needed.

Exercise 3.7. Generalize the previous argument to show that $W(3,r)$ exists for any $r \geq 2$. Can you derive a bound for $W(3,r)$?

Exercise 3.8. Sketch a proof that $W(4,3)$ exists by combining the arguments for $W(3,3)$ and $W(4,2)$. A careful drawing can represent most of the argument. What should the initial block size be in this case? The existence of which numbers $W(k,r)$ do we have to assume?

The general case.

Let us now try to generalize the previous arguments to arbitrary k,r. The idea remains the same: Find $(k-1)$-APs of larger and larger block patterns, until we can construct a number that is the "focal point" of r-many $(k-1)$-APs, each of a different color.

Formally, we will define a function $U(k,r)$ by recursion, which establishes an upper bound on $W(k,r)$, using van der Waerden's technique. Keep in mind that the actual van der Waerden numbers can be quite a bit smaller, as we have seen in the case of $k = 3$ and $r = 2$: The actual value is $W(3,2) = 9$, while $U(3,2) = 325$.

We start with the "innermost" block. For 3-APs, that block size was determined by taking $r + 1 = W(2,r)$ (i.e. the smallest number

3.1. Van der Waerden's theorem

for which a monochromatic 2-AP exists) and extending it sufficiently so that we can extend our 2-AP to a 3-AP. In the worst case, our 2-AP is given by the numbers 1 and $r+1$, and hence its extension to a 3-AP would be

$$1, \quad r+1, \quad 2r+1,$$

which means that the first box size is $2r + 1$. In the general case $U(k,r)$, we start with a $(k-1)$-AP over r colors, i.e. the interval $\{1, \ldots, U(k-1,r)\}$, and have to extend it by

$$\left\lfloor \frac{U(k-1,r)-1}{k-2} \right\rfloor.$$

Hence the block size for the innermost block would be

(3.4) $$b_1 = U(k-1,r) + \left\lfloor \frac{U(k-1,r)-1}{k-2} \right\rfloor.$$

Next, we consider coloring patterns of blocks of size b_1. There are r^{b_1} possibilities for each block, so to find a $(k-1)$-AP of b_1-blocks with the same coloring pattern we need $U(k-1, r^{b_1})$ of them. By putting

$$b_2 = U(k-1, r^{b_1}) + \left\lfloor \frac{U(k-1, r^{b_1})-1}{k-2} \right\rfloor,$$

we guarantee that among b_2 blocks of size b_1, there are k blocks in arithmetic progression, the first $k-1$ blocks of which have the same coloring pattern.

We can now continue inductively. Let

(3.5) $$b_{j+1} = U(k-1, r^{b_j}) + \left\lfloor \frac{U(k-1, r^{b_j})-1}{k-2} \right\rfloor,$$

and put

(3.6) $$U(k,r) = b_r \cdots b_2 b_1.$$

Then van der Waerden's argument yields that

$$W(k,r) \text{ exists and is at most } U(k,r).$$

It hopefully is clear why the argument (and the numbers) work. A formal proof is a little tedious—a double induction on k and r: Fixing k, we prove the result (inductively) for all r, and then use this (as we do in the definition of the numbers $U(k,r))$ to prove $W(k+1,2) \leq U(k+1,2)$.

Arguably the shortest and most elegant proof along the lines of van der Waerden's original ideas was given by Graham and Rothschild [23]. Another good presentation (with weaker but simpler bounds) can be found in the book by Khinchin [41].

A density version of van der Waerden's theorem. Turán's theorem shows that if we assume a certain density of edges, a complete subgraph of a certain size must be present. One can ask a similar question for van der Waerden's theorem: If a color is present with a certain density, do arithmetic progressions in that color exist? The answer is yes, and the result is known as **Szemerédi's theorem**, arguably one of the great results of 20th century mathematics.

Theorem 3.9 (Szemerédi's theorem [64]). *Let* $A \subset \mathbb{N}$. *If*

$$\limsup_{n \to \infty} \frac{|A \cap \{1, 2, \ldots, n\}|}{n} > 0,$$

then for any $k \geq 1$, A *contains infinitely many arithmetic progressions of length* k.

We will not prove Szemerédi's theorem here, but we wanted to at least state it, as the result has inspired some truly important developments in mathematics, from Furstenberg's ergodic-theoretic approach [17] to Gowers' uniformity norms [22] to the Green-Tao theorem on arithmetic progressions in the primes [26].

3.2. Growth of van der Waerden bounds

How good is the upper bound $U(k,r)$? It turned out not to be very good in the end, but it took quite a long time until significantly better bounds were discovered. And to this day, only slightly more than a handful of actual van der Waerden numbers are known for $k \geq 3$. We seem to encounter a phenomenon similar to what we saw for the Ramsey numbers: They grow fast and are notoriously difficult to compute. But in a certain sense, van der Waerden numbers (or rather, our bounds $U(k,r)$) are taking this phenomenon to the next level.

3.2. Growth of van der Waerden bounds

If you tried Exercise 3.6, you have probably seen that our upper bound $U(3,3) = (2 \cdot 3^{7 \cdot (2 \cdot 3^7 + 1)} + 1) \cdot (2 \cdot 3^7 + 1) \cdot 7$ is huge compared to $U(3,2) = 325$. You can probably imagine what will happen for $U(3,4)$.

Let us try to compute $U(4,2)$, using the definitions given by formulas (3.4), (3.5), and (3.6).

The innermost block size is given by

$$b_1 = U(3,2) + \left\lfloor \frac{U(3,2) - 1}{2} \right\rfloor = 325 + \frac{324}{2} = 487.$$

Then

$$b_2 = \left(U(3, 2^{487}) + \left\lfloor \frac{U(3, 2^{487}) - 1}{2} \right\rfloor \right) \cdot 487.$$

Hence to compute $U(4,2)$, we will need to know $U(3, 2^{487})$. In other words, we want a 3-AP for a 2^{487}-coloring. We have seen that $U(3,3)$ is already huge, but $U(3, 2^{487})$ seems truly astronomical (in fact, both numbers are already way, way larger than the number of atoms in the universe).

What is responsible for this explosive growth? Van der Waerden's argument uses a particular kind of recursion that generates such behavior. Incidentally, at about the same time van der Waerden proved the existence of monochromatic algebraic progressions, this kind of recursion came to prominence in the study of the foundations of mathematics.

The Ackermann function. We have seen in Chapter 1 that the diagonal Ramsey numbers $R(k)$ are bounded by 2^{2k}. Although we do not know an exact expression for $R(k)$, we say that the growth of the function $R(k)$ is at most exponential, as it is eventually dominated by a function of the form 2^{ck} for some constant c. A function $f : \mathbb{N} \to \mathbb{N}$ **eventually dominates** another function $g : \mathbb{N} \to \mathbb{N}$ if there exists a k_0 such that for all $k \geq k_0$, $f(k) \geq g(k)$. If we can choose $k_0 = 0$, i.e. if for all k, $f(k) \geq g(k)$, then we simply say that f **dominates** g.

What about the van der Waerden bound $U(k,r)$? As U is a binary function, we can measure its growth with respect to either variable, k or r, or look at the **diagonal** $U(m,m)$.

3. Growth of Ramsey functions

Let us first fix k and study the growth of $U(k,r)$ as a function of r. For $k = 2$, this is rather easy, since

$$U(2,r) = W(2,r) = r + 1;$$

in other words, $U(2,r)$ is of linear growth. Moving on to $k = 3$, recall from (3.6) that

$$U(k,r) = b_1 \cdots b_r,$$

where

$$b_1 = U(k-1,r) + \left\lfloor \frac{U(k-1,r) - 1}{k-2} \right\rfloor,$$

$$b_{j+1} = U(k-1, r^{b_j}) + \left\lfloor \frac{U(k-1, r^{b_j}) - 1}{k-2} \right\rfloor.$$

As we are interested in a lower bound for these values, we consider

$$b_1^* = U(k-1,r) \leq b_1,$$

$$b_{j+1}^* = U(k-1, r^{b_j^*}) \leq b_{j+1}.$$

This will make the computations that follow a little easier.

For $k = 3$, since $U(2,r) = r + 1$, we then have

$$b_1^* \geq r, \quad b_{j+1}^* \geq r^{b_j^*} \geq \underbrace{r^{r^{\cdot^{\cdot^{\cdot^r}}}}}_{(j+1) \text{ copies of } r}.$$

This means that

$$U(3,r) \geq b_r \geq b_r^* \geq \underbrace{r^{r^{\cdot^{\cdot^{\cdot^r}}}}}_{r \text{ copies of } r}.$$

This is a function that is not eventually dominated by any exponential function 2^{cr}, nor by a double exponential function $2^{2^{cr}}$, nor, in fact, by any finite-order exponential function $2^{2^{\cdot^{\cdot^{2^{cr}}}}}$, as the length of the exponential tower grows with the argument r. This "tower" operation

$$x \longmapsto \underbrace{x^{x^{\cdot^{\cdot^{\cdot^x}}}}}_{x \text{ copies of } x}$$

is called *tetration*. It is not very common, as we rarely encounter processes in mathematical or scientific practice that exhibit this kind of growth. This is arguably also the reason why we do not have a well-known notation for it (as we have for addition or exponentiation, for

3.2. Growth of van der Waerden bounds

example). For the van der Waerden bounds, however, tetration is just the beginning.

The important fact for us is that tetration can be defined recursively from exponentiation. What we mean by this is that we can set a ground value and then inductively define higher values by iterating exponentiation.

For example, multiplication is defined by iterating addition:

$$x \cdot 0 = 0,$$
$$x \cdot (y + 1) = x + x \cdot y.$$

And exponentiation is defined by multiplication:

$$x^0 = 1,$$
$$x^{y+1} = x \cdot (x^y).$$

Provided we know how to multiply two numbers, this gives us a recipe for computing the exponential of any pair of numbers, by calling the multiplication routine recursively. For example,

$$5^3 = 5 \cdot 5^2 = 5 \cdot (5 \cdot 5^1) = 5 \cdot (5 \cdot (5 \cdot (5^0)))$$
$$= 5 \cdot (5 \cdot (5 \cdot (1))) = 125.$$

In the first line we expand the definition backwards until we reach 0 in the exponent, for which we have a preset value, and then we substitute the value into the second line and evaluate forward.

In the same way we can define tetration from exponentiation. Let us use, as suggested by Knuth [**42**], the symbol $\uparrow\uparrow$:

$$x \uparrow\uparrow 0 := 1,$$
$$x \uparrow\uparrow (y+1) := x^{(x \uparrow\uparrow y)}.$$

Tetration is the fourth operation we obtain using the recursive iteration scheme, starting with addition, hence the name *tetra*, Greek for *four*.

If you unravel the definition, you will see that tetration indeed results in a *tower* of exponentiation

$$x \uparrow\uparrow y = \underbrace{x^{x^{\cdot^{\cdot^{\cdot^{x}}}}}}_{y \text{ times}}.$$

Our simple computation above shows that $U_3(r) := U(3, r)$ dominates $r \uparrow\uparrow r$. As $U(k+1, r)$ is defined by iterating $U(k, r)$, we would expect that $U(4, r)$ grows even faster, at least as fast as the iteration of $\uparrow\uparrow$. This is indeed the case, and to discuss this we introduce a general framework.

In 1928, the German mathematician Wilhelm Ackermann [1] defined a function $\varphi(x, y, n)$ that captured the idea of creating a new operation by iterating the previous one. Think of the third input n as an indicator for the level of the binary operation applied to the first two inputs.

Definition 3.10. The (3-place) **Ackermann function** φ is defined via the recursion

$$\varphi(x, y, 0) = x + y,$$

$$\varphi(x, 0, n+1) = \begin{cases} 0 & \text{if } n = 0, \\ 1 & \text{if } n = 1, \\ x & \text{if } n \geq 2, \end{cases}$$

$$\varphi(x, y+1, n+1) = \varphi(x, \varphi(x, y, n+1), n).$$

Anchoring the recursion is a bit more complicated because the number 0 behaves differently with respect to addition, multiplication, and exponentiation. But of course the last line contains the key idea: One iteration step at the $(n+1)$st level consists of applying the nth level-operation to a and the current iteration result at the $(n+1)$st level, $\varphi(a, b, n+1)$.

It is not completely clear that φ is well-defined. The recursion could be circular. One can show by double induction (the main induction on x, the side induction on y) that this is not the case, i.e. that by expanding the right-hand side of the recursion a finite number of times, we can express $\varphi(x, y, n)$ as an arithmetic combination of base

3.2. Growth of van der Waerden bounds

terms. The function φ is now known as the *Ackermann function*, although most texts usually present a binary variant due to Peter [50]. Let us define the nth **level of the Ackermann function** as

$$\varphi_n(x, y) := \varphi(x, y, n).$$

The recursion becomes

$$\varphi_{n+1}(x, y+1) = \varphi_n(x, \varphi_{n+1}(x, y)).$$

The function φ_n is more or less the same as the binary operation $\uparrow^{(n-1)}$ defined later by Knuth [42] (by, as you may guess, iterating the $\uparrow\uparrow$ function).

Exercise 3.11. Show that

$$\varphi_1(x, y) = x \cdot y,$$
$$\varphi_2(x, y) = x^y,$$
$$\varphi_3(x, y) = x \uparrow\uparrow (y+1).$$

Exercise 3.12. Show that the Ackermann function φ is monotone in all three places:

$$\varphi(x_0, y, n) \leq \varphi(x_1, y, n) \quad \text{whenever } x_0 \leq x_1,$$
$$\varphi(x, y_0, n) \leq \varphi(x, y_1, n) \quad \text{whenever } y_0 \leq y_1,$$
$$\varphi(x, y, m) \leq \varphi(x, y, n) \quad \text{whenever } 2 \leq m \leq n.$$

For $n \geq 3$, the functions φ_n grow extremely fast, even for small values. Let us compute $\varphi_4(2, 3)$.

3. Growth of Ramsey functions

Example 3.13.

$$\varphi_4(2,3) = \varphi_3(2,\varphi_4(2,2))$$
$$= \varphi_3(2,\varphi_3(2,\varphi_4(2,1)))$$
$$= \varphi_3(2,\varphi_3(2,\varphi_3(2,\varphi_4(2,0))))$$
$$= \varphi_3(2,\varphi_3(2,\varphi_3(2,2)))$$
$$= \varphi_3(2,\varphi_3(2,2\uparrow\uparrow 2))$$
$$= \varphi_3(2,\varphi_3(2,2^2))$$
$$= \varphi_3(2, 2\uparrow\uparrow 4)$$
$$= \varphi_3(2, 2^{2^{2^2}})$$
$$= 2\uparrow\uparrow 65536$$
$$= \underbrace{2^{\cdot^{\cdot^{\cdot^2}}}}_{65536 \text{ times}}$$

The van der Waerden bound $U(k,r)$ in turn, for fixed k and as a function of r, dominates the kth level of the Ackermann function. We have already seen this for $k = 3$. So let us assume we have shown that for $k \geq 3$ and $r \geq 2$, $U_k(r) \geq \varphi_k(r, r-1)$. Now

$$U(k+1, r) \geq b_1^* \cdots b_r^* = U(k,r)U(k, r^{b_1^*}) \cdots U(k, r^{b_{r-1}^*}),$$

and with a number of rather crude estimates, using the monotonicity of φ we obtain

$$U(k+1, r) \geq \varphi_k(r, r-1)\,\varphi_k(r^{b_1^*}, r^{b_1^*} - 1) \cdots \varphi_k(r^{b_{r-1}^*}, r^{b_{r-1}^*} - 1)$$
$$\geq \varphi_k(r^{b_{r-1}^*}, r^{b_{r-1}^*} - 1)$$
$$\geq \varphi_k(r, b_{r-1}^*)$$
$$\geq \varphi_k(r, \varphi_k(r^{b_{r-2}^*}, r^{b_{r-2}^*-1}))$$
$$\geq \varphi_k(r, \varphi_k(r, b_{r-2}^*))$$
$$\geq \underbrace{\varphi_k(r, \varphi_k(r, \ldots \varphi_k(r, b_1^*) \ldots))}_{(r-1) \text{ iterations}} \geq \varphi_{k+1}(r, r-1).$$

Therefore, we have the following.

Proposition 3.14. *For $k \geq 3$ and $r \geq 2$,*

$$U(k,r) \geq \varphi_k(r, r-1).$$

3.3. Hierarchies of growth

While, as we have seen, the growth of $\varphi_k(r,r)$ is already rather impressive for $k \geq 4$, truly "explosive" growth is generated when we allow k to vary, too.

Exercise 3.15. Show that $U_\omega(k) = U(k,k)$ eventually dominates $\varphi_m(k,k)$ for any m.

3.3. Hierarchies of growth

In the previous section we introduced the Ackermann function φ. We saw that the associated *level functions*, φ_k, track the growth of the kth van der Waerden bound $U_k(r)$. In this section we will connect these functions to important concepts from computability theory, the notion of a *primitive recursive function* and the *Grzegorczyk hierarchy*. We will also see how we can generate even faster-growing functions, by extending the construction into the transfinite. This is where ordinals will be needed again.

Primitive recursive functions. When we introduced the Ackermann function φ (Definition 3.10), we observed that each level φ_k is obtained by iterating the previous one—multiplication iterates addition, exponentiation iterates multiplication, and so on. We also defined an upper bound $U(k,r)$ for the van der Waerden number $W(k,r)$ using a similar iteration. The actual definition of $U(k,r)$ was a bit more complicated than that of φ, but essentially we used two **basic operations** to define $U(k,r)$:

Composition: Given functions $h(x_1, \ldots, x_m)$ and
$g_1(x_1, \ldots, x_n), \ldots, g_m(x_1, \ldots, x_n)$, we can compose these functions to define a new function
$$f(x_1, \ldots, x_m) = h(g_1(x_1, \ldots, x_n), \ldots, g_m(x_1, \ldots, x_n)).$$

Recursion: Given functions $g(x_1, \ldots, x_n)$ and $h(x_1, \ldots, x_n, y, z)$, we can define f by recursion from g and h as
$$f(x_1, \ldots, x_n, 0) = g(x_1, \ldots, x_n),$$
$$f(x_1, \ldots, x_n, y+1) = h(x_1, \ldots, x_n, y, f(x_1, \ldots, x_n, y)).$$

The arities of the functions considered here are arbitrary but have to be finite.

Each function φ_k, as well as each kth level of the van der Waerden bound U_k, can be obtained by a finite number of applications of these two operations to a set of basic functions: $x+y$ in the case of φ, $r+1$ (the value of $W(2,r)$), multiplication, exponentiation, and integer division for U.

Functions that can be obtained this way belong to an important family of functions—the primitive recursive functions. To define them for all finite arities, we introduce the following **basic functions**.

Zero function: $\text{Zero}(x) = 0$.

The successor function: $S(x) = x+1$.

The projection functions: $P_n^i(x_1, \ldots, x_n) = x_i$.

Definition 3.16. The family of **primitive recursive functions** is the smallest family of functions $f : \mathbb{N}^n \to \mathbb{N}$ (where $n \in \mathbb{N}$ can be arbitrary) that contains the three basic functions and is closed under composition and recursion.

In other words, a function is primitive recursive if it can be obtained from the basic functions by a finite number of applications of composition and recursion. For example, addition + is primitive recursive as it can be defined as follows:

$$x + 0 = P_1^1(x),$$
$$x + (y+1) = S(P_3^1(x, y, +(x,y))).$$

One can continue inductively and obtain the following.

Proposition 3.17. *Every level $\varphi_k(x,y)$ of the Ackermann function is primitive recursive.*

The family of primitive recursive functions includes many more functions. For example, the common functions of elementary number theory, such as gcd and remainder, are primitive recursive. The verification of this is often a little tedious. One has to build up a "library" of primitive recursive functions. We give some examples.

Lemma 3.18. *The predecessor function*

$$x \dotminus 1 = \begin{cases} x-1 & \text{if } x > 0, \\ 0 & \text{if } x = 0 \end{cases}$$

is primitive recursive.

3.3. Hierarchies of growth

Proof. We can define this recursively as

$$0 \dotdiv 1 = 0 = \text{Zero}(0),$$
$$(x+1) \dotdiv 1 = x = (x \dotdiv 1) + 1 = S(x \dotdiv 1).$$

□

Exercise 3.19. Show that the following functions are primitive recursive:

$$x \dotdiv y, \quad \max(x,y), \quad \min(x,y), \quad |x-y|.$$

An important property of the primitive recursive functions is their closure under *bounded* search procedures.

Definition 3.20. Given a predicate $P(\vec{x}, y)$ on $\mathbb{N}^{n+1}$, the **bounded μ-operator** applied to a predicate $P(\vec{x}, y)$ is defined as

$$\mu z < y \, [P(\vec{x}, z)] = \begin{cases} z & \text{if } z \text{ is the least natural number} < y, \\ & \text{such that } P(\vec{x}, y) \text{ holds} \\ y & \text{if no such } z \text{ exists.} \end{cases}$$

This means that the bounded μ-operator applied to a predicate returns the least witness less than y such that the predicate holds.

Proposition 3.21. *If $g(\vec{x}, y)$ is primitive recursive, then so is*

$$f(\vec{x}, y) = \mu z < y \, [g(\vec{x}, z) = 0].$$

For a proof see, for example, [**54**, Section 6.1].

Exercise 3.22. Use the closure under the bounded μ-operator to show that the following functions are primitive recursive:

$$\text{div}(x, y) = \text{result of integer division of } x \text{ by } y,$$
$$\text{rem}(x, y) = \text{remainder of integer division of } x \text{ by } y.$$

In the late 19th century, work by Dedekind [**11**] and Peano [**48, 49**] demonstrated that definitions by induction play an important role in the development of number theory from first principles (axioms), and that any axiomatic framework for number theory must include a principle ensuring that functions described by induction (recursion) are indeed well-defined. This led to the introduction of the family of primitive recursive functions as the closure of the basic arithmetic functions under composition and inductive definitions.

Furthermore, every primitive recursive function is *computable* (in an intuitive sense), as we can compute any value using a finite deterministic procedure, unraveling the recursion, as we did in Example 3.13. Moreover, all the usual number-theoretic functions, which are intuitively computable, can also be shown to be primitive recursive.

The Grzegorczyk hierachy. The Ackermann function provided the first example of a computable function (in the intuitive as well as in a formal sense) that was *not* primitive recursive. The key here is that the diagonal Ackermann function eventually dominates every primitive recursive function. In fact, the levels φ_k of the Ackermann function form a kind of "spine" of the primitive recursive functions, in terms of their growth behavior. To formulate this, it will be useful to have a unary version of the Ackermann level functions.

Definition 3.23. The family of functions $\{\Phi_n : n \in \mathbb{N}\}$ is defined by

$$\Phi_n(x) = \begin{cases} 2 & \text{if } n = 0, \\ \varphi_n(2, x) & \text{if } n > 0. \end{cases}$$

Hence we have

$$\Phi_0(x) = x + 2,$$
$$\Phi_1(x) = 2x,$$
$$\Phi_2(x) = 2^x,$$
$$\Phi_3(x) = 2 \uparrow\uparrow (x+1),$$

and in general, for $n \geq 1$,

$$\Phi_{n+1}(0) = 2,$$
$$\Phi_{n+1}(x+1) = \Phi_n(\Phi_{n+1}(x)).$$

In other words, Φ_{n+1} is obtained by iterating Φ_n. If we denote the n-iteration of a function f by $f^{(n)}$, then

$$\Phi_{n+1}(x) = \Phi_n^{(x)}(2).$$

3.3. Hierarchies of growth

We already observed that by iterating a function we are able to generate another function that grows much faster than the original function. This can be made precise by defining a hierarchy on the family of primitive recursive functions that categorizes them according to their growth behavior. Such a hierarchy was introduced by Grzegorczyk [28] and has since become known as the **Grzegorczyk hierarchy**

$$\mathcal{E}_0 \subset \mathcal{E}_1 \subset \mathcal{E}_2 \subset \cdots.$$

$\mathcal{E}_0$ is the smallest family of functions containing the base functions Zero, $x + y$ (instead of successor), and P_n^i and that is closed under composition and **limited recursion** with respect to $\mathcal{E}_0$. Here limited recursion means that if f is defined via recursion from $g, h \in \mathcal{E}_0$ and there exists a function $j \in \mathcal{E}_0$ such that for all $x_1, \ldots, x_n, y \in \mathbb{N}$,

$$f(x_1, \ldots, x_n, y) \leq j(x_1, \ldots, x_n, y),$$

then $f \in \mathcal{E}_0$. In other words, to add a function to $\mathcal{E}_0$, the growth of the function has to be bounded by a witness j already known to be in $\mathcal{E}_0$.

Once $\mathcal{E}_n$ is defined, $\mathcal{E}_{n+1}$ is defined as the smallest family of functions containing all functions from $\mathcal{E}_n$, the function Φ_n (or rather, a slight variant thereof) and which is closed under composition and limited recursion.

Defining $\mathcal{E}_n$ via a *closure property* ensures that each $\mathcal{E}_n$ is very robust with respect to operations of its *own type*. For example, $\mathcal{E}_2$ contains the multiplication function, and if we multiply two functions from $\mathcal{E}_2$, we obtain a function in $\mathcal{E}_2$ again. On the other hand, if we iterate a function from $\mathcal{E}_n$, we obtain a function in $\mathcal{E}_{n+1}$. More generally, if $g, h \in \mathcal{E}_n$ and f is defined from g and h via recursion, then $f \in \mathcal{E}_{n+1}$. Moreover, for all n, $\Phi_n \in \mathcal{E}_{n+1} \setminus \mathcal{E}_n$. In other words, the hierarchy is *proper*.

The Grzegorczyk hierarchy captures all primitive recursive functions with respect to their growth behavior in the following sense.

Theorem 3.24.

(a) *Every primitive recursive function is contained in some $\mathcal{E}_n$.*

(b) *If f is a unary function in $\mathcal{E}_n$, then f is eventually dominated by Φ_n.*

It would go beyond the scope of this book to prove the theorem here, so we refer to the book by Rose [56].

Corollary 3.25. *The diagonal function $\Phi_\omega(x) = \Phi_x(x)$ is not primitive recursive.*

Proof. If Φ_ω were primitive recursive, there would exist k and x_0 such that for all $x > x_0$,

$$\Phi_\omega(x) \le \Phi_k(x).$$

But for $x > \max(k, x_0)$, we have by the monotonicity of the Φ_n that

$$\Phi_k(x) < \Phi_x(x) = \Phi_\omega(x),$$

which is a contradiction. $\square$

Corollary 3.26. *The Ackermann function $\varphi(x, y, n)$ and the van der Waerden upper bound $U(k, r)$ are not primitive recursive.*

A word of caution: You may be tempted to think now that if a function is bounded by some Φ_n, it is primitive recursive. However, this is far from the truth. In Chapter 4, we will encounter a $\{0, 1\}$-valued function that is not even (Turing) computable, let alone primitive recursive.

Extending the hierarchy. The functions Φ_n and *diagonal* function $\Phi_\omega(x) = \Phi_x(x)$ give us a blueprint of how to construct ever faster-growing functions.

We iterated Φ_n to create the faster-growing function Φ_{n+1}. Let us apply this now to the diagonal function Φ_ω:

$$\Phi_{\omega+1}(0) = 2,$$
$$\Phi_{\omega+1}(x+1) = \Phi_\omega(\Phi_{\omega+1}(x)).$$

As you can probably tell, we have chosen the index ω with the idea in mind that the ordinals will help us index our extended hierarchy

3.3. Hierarchies of growth

beyond the finite levels. So, in general, assume that we have defined the function $\Phi_\alpha(x)$ for some ordinal α, and let

$$\Phi_{\alpha+1}(0) = 2,$$
$$\Phi_{\alpha+1}(x+1) = \Phi_\alpha(\Phi_{\alpha+1}(x)).$$

Note that even though the functions are indexed by ordinals, they are still functions from $\mathbb{N}$ to $\mathbb{N}$.

What this definition does not tell us is how to define Φ_α in the case where α is a limit ordinal. For $\alpha = \omega$, we took the diagonal of the functions "leading up to it". We can do something similar for an arbitrary limit ordinal α by putting

$$\Phi_\alpha(x) = \Phi_{\alpha_x}(x),$$

where α_x is a sequence of ordinals with limit α, i.e. $\sup\{\alpha_x : x \in \mathbb{N}\} = \alpha$. Any such sequence is called a **fundamental sequence for** α.

There are two issues: First, this only works for limit ordinals with cofinality ω (recall the definition of cofinality, Definition 2.43). As we are only interested in Φ_α for countable α, this does not really present a problem. Second, there are multiple ways to choose a fundamental sequence. For example, if $\alpha = \omega + \omega$, both

$$\omega + 1, \omega + 3, \omega + 5, \ldots \quad \text{and} \quad \omega + 2, \omega + 4, \omega + 6, \ldots$$

converge to α. Is there a *canonical way* of selecting a fundamental sequence? If we consider only $\alpha < \varepsilon_0$ (recall that ε_0 was the least ordinal ε for which $\omega^\varepsilon = \varepsilon$), there is indeed such a way, given through the **Cantor normal form**.

Theorem 3.27 (Cantor normal form). *Every ordinal $0 < \alpha < \varepsilon_0$ can be represented uniquely in the form*

$$\alpha = \omega^{\beta_1} + \cdots + \omega^{\beta_n},$$

where $n \geq 1$ and $\alpha > \beta_1 \geq \cdots \geq \beta_n$.

In fact, if we only require $\alpha \geq \beta_1$, every ordinal α has a unique representation of this form.

For example, $\alpha = \omega^{\omega^2+3} + \omega^2 + \omega^2 + 1 + 1 + 1$ is in Cantor normal form. (Recall that $\omega^0 = 1$.)

Proof. We proceed by induction. For $\alpha = 1$, we have $\alpha = \omega^0$. Suppose now that the assertion holds for all $\beta < \alpha$. Let
$$\gamma^* = \sup\{\gamma \colon \omega^\gamma \le \alpha\}.$$
Then γ^* is the largest ordinal such that $\omega^{\gamma^*} \le \alpha$, and as $\alpha < \varepsilon_0$, $\gamma^* < \alpha$.

Note that for any two ordinals $\delta \le \alpha$, there exists a unique ρ such that $\alpha = \delta + \rho$. We simply let ρ be the order type of the set $\{\beta \colon \delta < \beta \le \alpha\}$.

Now let ρ be the unique ordinal such that $\alpha = \omega^{\gamma^*} + \rho$. By the choice of γ^*, we must have $\rho < \alpha$ (see Exercise 3.28), and by the inductive hypothesis, ρ has a normal form
$$\rho = \omega^{\gamma_1} + \cdots + \omega^{\gamma_k}.$$
Then
$$\alpha = \omega^{\gamma^*} + \omega^{\gamma_1} + \cdots + \omega^{\gamma_k},$$
with $\gamma^* \ge \gamma_1 \ge \cdots \ge \gamma_k$, as desired. $\square$

Exercise 3.28. Show that $\omega^\gamma + \alpha = \alpha$ implies $\omega^{\gamma+1} \le \alpha$.

Exercise 3.29. Use transfinite induction to show that the Cantor normal form is unique.

How can we use the Cantor normal form in selecting a fundamental sequence for α?

Suppose we have
$$\alpha = \omega^{\beta_1} + \cdots + \omega^{\beta_n},$$
with $\beta_1 \ge \cdots \ge \beta_n$. If α is a limit ordinal, we must have $\beta_n \ge 1$. If β_n is a successor ordinal, i.e. if $\beta_n = \gamma + 1$, we can let
$$\alpha_k = \omega^{\beta_1} + \cdots + \omega^\gamma \cdot k,$$
and so $\lim_k \alpha_k = \alpha$. If β_n is a limit ordinal, we have $\beta_n < \alpha$ (since $\alpha < \varepsilon_0$), and we can construct a fundamental sequence by induction: Assume that we have constructed one for β_n, say $(\gamma_k)_{k \in \mathbb{N}}$; then $(\alpha_k)_{k \in \mathbb{N}}$ given by
$$\alpha_k = \omega^{\beta_1} + \cdots + \omega^{\beta_{n-1}} + \omega^{\gamma_k}$$
is a fundamental sequence for α.

3.4. The Hales-Jewett theorem

To make our definition of Φ_α rigorous (assuming $\alpha < \varepsilon_0$ is a limit ordinal), let us write $\alpha[k]$ for the kth term in the canonical fundamental sequence for α defined above. We put

$$\Phi_\alpha(x) = \Phi_{\alpha[x]}(x).$$

Finally we define $\beta_1 = \omega$ and $\beta_{n+1} = \omega^{\beta_n}$, that is, β_n is a tower of n-many ω's. Then let

$$\Phi_{\varepsilon_0}(x) = \Phi_{\beta_x}(x).$$

The functions Φ_α, $\alpha \leq \varepsilon_0$, are known as the **Wainer hierarchy**. They are still functions from $\mathbb{N}$ to $\mathbb{N}$. We can even write a computer program to evaluate them. This seems strange at first sight, as the ordinals used to index them are infinite. But they are defined via recursion, and there are no infinite descending chains of ordinals. Only finitely many steps of "unraveling" are required. Try to compute $\Phi_{\omega^2+17}(3)$, and you will get the idea. (Of course the running time of this algorithm on any computer would exceed the expected life of this computer by far.)

Still, the Φ_α, in particular Φ_{ε_0}, seem somewhat unreal, a pure construct based on the infinitude of ordinals. Are there any "real" mathematical objects tied to this kind of growth?

The remaining two sections of this chapter will go in opposite directions with respect to this question. First, we will bring the van der Waerden numbers to level Φ_4. After that, we will see that a rather simple modification of Ramsey's original theorem will produce a true growth behavior at the level of ε_0.

3.4. The Hales-Jewett theorem

While van der Waerden's theorem proved the existence of patterns in sequences of numbers, the theorem of Hales and Jewett [29], originally proven in 1963, deals with more general combinatorial geometric objects.

The **combinatorial cube** C_t^n is the set of n-tuples with entries in $[t] := \{1, \ldots, t\}$; each n-tuple is called a *combinatorial point*. One can picture this as an n-dimensional cubic array with side length t. An r-coloring on a combinatorial cube is then a function $c : C_t^n \to$

114   3. Growth of Ramsey functions

$[r]$. Each cell in the array is labeled with one of r colors. As an example, consider the game tic-tac-toe. A standard tic-tac-toe board is a representation of the combinatorial cube C_3^2, and a game of tic-tac-toe is essentially taking turns defining a 2-coloring on the cube. With the game of tic-tac-toe in mind, it makes sense to ask about the existence of monochromatic lines.

Definition 3.30. For $n, t > 1$, a **combinatorial line** in C_t^n is a subset of t distinct combinatorial points $L_1, \ldots, L_t$ where, for each $1 \le i \le n$, either

$$L_{\tau,i} = L_{\sigma,i} \text{ for all } 1 \le \tau, \sigma \le t,$$

that is, the ith coordinate is constant, or

$$L_{\tau,i} = \tau \text{ for all } 1 \le \tau \le t,$$

which means that the ith coordinate varies with τ.

Note that in a combinatorial line, at least one coordinate has to vary with τ. An example of a combinatorial line in C_5^4 is

$$L_1 = (1\ 4\ 1\ 2),$$
$$L_2 = (2\ 4\ 2\ 2),$$
$$L_3 = (3\ 4\ 3\ 2),$$
$$L_4 = (4\ 4\ 4\ 2),$$
$$L_5 = (5\ 4\ 5\ 2).$$

The notion of a combinatorial line is slightly stricter than for instance the lines allowed in tic-tac-toe, because the diagonal from the upper left to the lower right corner on a tic-tac-toe board,

$$(1\ 3)(2\ 2)(3\ 1),$$

is not a combinatorial line, while the lower left to upper right diagonal

$$(1\ 1)(2\ 2)(3\ 3)$$

is.

Exercise 3.31. Determine all combinatorial lines for C_4^2. How many are there?

3.4. The Hales-Jewett theorem

We will often use an alternative notation for combinatorial lines. A (t,n)-*-**word** is a sequence over $\{1,\ldots,t\}\cup\{*\}$ of length n where at least one entry is a *. Any (t,n)-*-word L represents a combinatorial line in C_t^n by simultaneously substituting τ for every occurrence of * in L, for $1 \le \tau \le t$. We denote the resulting point in C_t^n by $L(\tau)$. The first example above corresponds to the *-word $L = (*\ 4\ *\ 2)$, while the diagonal on a tic-tac-toe board is given by $(*\ *)$.

Exercise 3.32. Represent the combinatorial lines in C_4^2 using *-words. Derive a general formula for the number of combinatorial lines in C_t^n.

Theorem 3.33 (Hales and Jewett). *For any integers $t, r > 1$, there exists an integer $HJ(t,r)$ such that for any $n \ge HJ(t,r)$, every r-coloring of C_t^n has a monochromatic combinatorial line.*

The integers $HJ(r,t)$ are called the **Hales-Jewett numbers**. It is well known that the standard tic-tac-toe is a boring game; with optimal strategies, the game will always end in a tie. This means that there exist (lots of) 2-colorings of C_3^2 which contain no monochromatic combinatorial lines, and therefore $HJ(3,2) > 2$. In 2014, Hindman and Tressler [**33**] proved that $HJ(3,2) = 4$. We should probably play tic-tac-toe in four-dimensions.

The rather geometric concept of combinatorial lines is very versatile. Many combinatorial statements can be transformed into an equivalent statement about combinatorial lines. This makes the Hales-Jewett theorem extremely powerful. We illustrate this by deriving van der Waerden's theorem from it.

Proposition 3.34. *If $HJ(t,r)$ exists, then $W(t,r)$ exists and it holds that $W(t,r) \le t^{HJ(t,r)}$, where $W(t,r)$ is the van der Waerden number (defined in Theorem 3.3).*

Proof. Consider the integers from $\{0,\ldots,t^n-1\}$ in their base-t representation:
$$x = x_0 + x_1 t + \cdots + x_n t^{n-1},$$
where $0 \le x_i \le t-1$. We can identify x with $(x_0, x_1, \ldots, x_n)$, and this provides a bijection $p : \{0, \ldots, t^n - 1\} \to C_t^n$. (Note that we are slightly

deviating from our previous notation by having the coordinates range from 0 to $t-1$ rather than from 1 to t.) Then, any r-coloring on $[0, t^n - 1]$ induces an r coloring on C_t^n.

If we assume that the Hales-Jewett theorem is true and that $n > HJ(t, r)$, then C_t^n has a monochromatic line, say L. Let A be the subset of $\{0, \ldots, n-1\}$ for which the entries of L are a $*$ and let B be $\{0, \ldots, n-1\} \setminus A$. Then the preimage in $\{0, \ldots, t^n - 1\}$ of any point on the line L is of the form

$$x = p^{-1}(L(\tau)) = \sum_{i=1}^{n} L(\tau)_i \, t^{i-1} = \sum_{i \in A} \tau t^{i-1} + \sum_{i \in B} L_i t^{i-1}.$$

If we let $D = \sum_{i \in A} t^{i-1}$ and $C = \sum_{i \in B} L_i t^{i-1}$, then $x = D\tau + C$, and therefore the preimage of L forms a monochromatic arithmetic progression in $\{0, \ldots, t^{n-1}\}$ of length t. $\square$

Exercise 3.35. Determine the arithmetic progression in $\mathbb{N}$ induced by $L = (* \, 3 \, *)$ in C_4^3.

The original proof by Hales and Jewett used a double induction and hence gave bounds that, similar to van der Waerden's theorem, grow in an Ackermann-like fashion. In 1988, Shelah [**58**] gave a completely new proof that brought the growth of the Hales-Jewett numbers (and hence, by Proposition 3.34, also the Van der Waerden numbers) "down to Earth", by establishing the existence of $HJ(t, r)$ as a single induction on t (with r arbitrary but fixed).

The proof of the Hale-Jewett theorem given here follows the one by Shelah, where we adopt some of the organizational structure (and notation) from the presentation of [**24**].

Shelah was able to reduce the problem for r-colorings of C_t^n to r-colorings of C_{t-1}^n. In fact, if the given coloring is not able to distinguish between $t-1$ and t, we can extend a monochromatic line of length $t-1$ to one of length t.

To be specific, two points $x = (x_1, \ldots, x_n)$ and $y = (y_1, \ldots, y_n)$ in C_t^n are t-**close** if for all $i \leq n$,

(a) $x_i < t - 1$ if and only if $y_i < t - 1$, and

(b) $x_i < t - 1$ implies $x_i = y_i$.

3.4. The Hales-Jewett theorem

In other words, x and y agree on entries below $t-1$ and can differ by at most 1 on the other entries. For example,

$$(1\ 2\ 4\ 4\ 1\ 3) \text{ and } (1\ 2\ 3\ 4\ 1\ 4)$$

are 4-close points in C_4^6, whereas

$$(1\ 2\ 4\ 2\ 1\ 3) \text{ and } (4\ 2\ 3\ 4\ 1\ 4)$$

are not.

An r-coloring c on C_t^n is t-**blind** if any pair of t-close points has the same color.

Lemma 3.36. *If $HJ(t-1,r)$ exists, $n \geq HJ(t-1,r)$, and c is a t-blind r-coloring on C_t^n, then there exists a monochromatic line.*

Proof. Restrict c to a coloring on $C_{t-1}^n \subset C_t^n$. By assumption, there is a monochromatic line L in C_{t-1}^n. This line is a set of $t-1$ points and is a subset of a unique line L^+ in C_t^n having one additional point which is equal to $L(t)$. We have the following:

(a) $c(L(1)) = c(L(2)) = \cdots = c(L(t-1))$, because L is monochromatic in C_{t-1}^n;

(b) $c(L(t-1)) = c(L(t))$, because $L(t-1)$ and $L(t)$ are t-close and the coloring is t-blind.

Therefore, the entire line L is monochromatic in C_t^n. □

Of course not every coloring of C_t^n is t-blind. But Shelah managed to show that for any coloring on a sufficiently large cube, there is a subset of a combinatorial substructure where the coloring *is* t-blind. In the following, if a coloring restricted to a combinatorial structure is t-blind, we will call this structure t-blind, too. For example, we will speak of *blind combinatorial lines*.

Shelah's combinatorial substructures are built from rather simple blocks. For $1 \leq a \leq b \leq n$, define a **Shelah line**[2] $L^{a,b}$ in C_t^n as the $*$-word of the form

$$(\underbrace{t-1 \cdots t-1}_{(a-1)\text{-many}}\ \underbrace{* \cdots *}_{(b-a+1)\text{-many}}\ \underbrace{t \cdots t}_{(n-b)\text{-many}})$$

---

[2] The naming of these combinatorial structures after Shelah is due to Graham, Rothschild, and Spencer [**24**], and we follow them here.

In C_4^6, the Shelah line $L^{3,5}$ is the $*$-word (3 3 $*$ $*$ $*$ 4), which induces the line

$$(3\ 3\ 1\ 1\ 1\ 4),$$
$$(3\ 3\ 2\ 2\ 2\ 4),$$
$$(3\ 3\ 3\ 3\ 3\ 4),$$
$$(3\ 3\ 4\ 4\ 4\ 4).$$

Let us call a point on a Shelah line a **Shelah point**. Such a point depends only on a, b, and the value τ of $*$. Therefore, there are no more than $t \cdot n^2$ Shelah points in C_t^n.[3]

Lemma 3.37 (Shelah line pigeonhole principle). *If $n \geq r$, then for any r-coloring c of C_t^n, there exists a Shelah line restricted to which c is t-blind.*

Proof. For $0 \leq i \leq n$, let $P_i \in C_t^n$ be the point where the first i coordinates are equal to $t-1$ and the last $n-i$ coordinates are equal to t. There are $n+1$ such points, and each has been colored with one of r colors, so by the pigeonhole principle, at least two such points have the same color, say P_i and P_j. Note that these are both Shelah points on the same line with $P_j = L^{i+1,j}(t-1)$ and $P_i = L^{i+1,j}(t)$, so $c(L_{i+1,j}(t-1)) = c(L_{i+1,j}(t))$ and therefore the Shelah line $L^{i+1,j}$ is t-blind. □

A single Shelah line will not be enough for our purposes; we need the simultaneous existence of multiple Shelah lines.

A **combinatorial s-space**, Σ_s, is the concatenation of s combinatorial lines of variable dimensions over the same alphabet. We can represent the space as the set of s-tuples whose entries are points on combinatorial lines,

$$(L^1(\tau_1), L^2(\tau_2), \ldots, L^s(\tau_s)).$$

Each L^i is a combinatorial line in $C_t^{n_i}$ for some n_i, and each τ_i can vary independently. The combinatorial s-space Σ_s can be realized as a subset of C_t^n where $n = \sum n_i$.

---

[3]This bound is clearly not optimal, but it facilitates counting later while having little impact on the overall growth estimate.

3.4. The Hales-Jewett theorem

A **Shelah s-space** is a combinatorial space whose coordinates are all Shelah lines. A Shelah line has one degree of freedom (the value of the $*$-block), so there is a canonical bijection between a Shelah s-space S and C_t^s. Let us denote this bijection by $\pi : C_t^s \to S$. For example, consider the Shelah 3-space

$$(2\,2\, * \,3) \times (2\, *) \times (2\, *\, *\, *\, 3\,3).$$

While points in this space live in C_3^{12}, any such point is completely determined by the respective values of $*$ in the three Shelah lines. Therefore, there is a canonical bijection between the space and C_3^3.

Given a coloring c of a Shelah s-space, we call c t-blind if the induced coloring $c^* = c \circ \pi$ is t-blind.

Lemma 3.38 (Existence of blind Shelah spaces). *Given integers $r, s, t \geq 1$, there exists an N such that every r-coloring of C_t^N contains a Shelah s-space on which the coloring is t-blind.*

Proof. Suppose a sufficiently large N is given and we want to find a partition

$$N = N_1 + \cdots + N_s$$

such that there exists a blind Shelah space the lines of which lie in $C_t^{N_i}$. We can define the N_i and the corresponding Shelah lines via reverse induction: Starting with N_s, at each step i we determine how large to choose N_i in relation to the preceding N_j, $j < i$, so that we can not only find a t-blind Shelah line in $C_t^{N_i}$ but also leave enough points in the preceding $C_t^{N_j}$ color compatible with the Shelah line in $C_t^{N_i}$. The main tool for this will be an iterated application of Lemma 3.37, the Shelah line pigeonhole principle.

Let us assume $N = N_1 + \cdots + N_s$ and thus

$$(C_t^{N_1}, \ldots, C_t^{N_s}) \cong C_t^N.$$

Given a coloring c on C_t^N, we define an equivalence relation on $C_t^{N_s}$ as follows: For two points $y, y^* \in C_t^{N_s}$, we will say that y and y^* are *equivalent* if for all Shelah points x_i in $C_t^{N_i}$ with $i < s$,

$$c(x_1, \ldots, x_{s-1}, y) = c(x_1, \ldots, x_{s-1}, y^*).$$

3. Growth of Ramsey functions

Two points are hence equivalent if they have identical coloring behavior with respect to all $(s-1)$-combinations of Shelah point predecessors.

Since each $C_t^{N_i}$ has no more than tN_i^2 Shelah points, each point in $C_t^{N_s}$ has at most

$$t^{s-1} \prod_{j=1}^{s-1} N_j^2 =: M_s$$

such Shelah $(s{-}1)$-predecessors. Moreover, each of these Shelah $(s{-}1)$-predecessors is colored one of r ways, so each point $y \in C^{N_s}$ has r^{M_s} possible coloring behaviors with respect to its Shelah $(s-1)$-predecessors. Therefore, there are at most r^{M_s} equivalence classes.[4]

Equivalence relations naturally correspond to colorings (by assigning each equivalence class a different color). Our equivalence relation thus defines an r^{M_s}-coloring on $C_t^{N_s}$. If we require $N_s \geq r^{M_s}$, we can apply the Shelah line pigeonhole principle (Lemma 3.37) and conclude that $C_t^{N_s}$ contains a t-blind Shelah line with respect to this new coloring, that is, a line L^s in $C_t^{N_s}$ where the points $L^s(t-1)$ and $L^s(t)$ are in the same equivalence class.

Now that we have found a t-blind Shelah line in $C_t^{N_s}$, we will work inductively backward to complete our t-blind Shelah space, while at the same time bounding the N_j in terms of their predecessors (and their predecessors *only*). Assume that we have found t-blind Shelah lines $L^{i+1}, \ldots, L^s$ in the last $s-i$ cubes $C_t^{N_{i+1}}, \ldots, C_t^{N_s}$.

We define an equivalence relation on $C_t^{N_i}$ similar to the one on $C_t^{N_s}$: Two points $y, y^* \in C_t^{N_i}$ are equivalent if

$$c(x_1, \ldots, x_{i-1}, y, z_{i+1}, \ldots, z_s) = c(x_1, \ldots, x_{i-1}, y^*, z_{i+1}, \ldots, z_s)$$

for all Shelah points $x_j \in C_t^{N_j}$, $j < i$, and all points $z_k \in L^k$, $k > i$. That is, the colorings agree with respect to all Shelah points in the preceding components and all points chosen from the Shelah lines already fixed in the subsequent components.

There are fewer equivalence classes now, though, because while we can choose any Shelah point from $C_t^{N_1}$ through $C_t^{N_{i-1}}$, we have

---

[4]This is similar to the block-coloring patterns in the proof of van der Waerden's theorem.

3.4. The Hales-Jewett theorem

restricted ourselves to only one of t Shelah points from $C_t^{N_{i+1}}$ through $C_t^{N_s}$. Therefore, there are at most

$$\prod_{j=1}^{i-1}(tN_j^2) \cdot t^{s-i} = t^{s-1}\prod_{j=1}^{i-1} N_j^2 =: M_i$$

possible choices for the x_j and z_k. As before, this means that there are r^{M_i} equivalence classes. We require $N_i \geq r^{M_i}$, and as before the Shelah line pigeonhole principle implies that $C_t^{N_j}$ contains a t-blind Shelah line, L^i, with respect to this equivalence relation.

Continuing in this way, we obtain Shelah lines $L^1, \ldots, L^s$ and claim that the original coloring c is t-blind for the corresponding Shelah s-space S.

It suffices to verify the following: If $(z_1, \ldots, z_{i-1}, y, z_{i+1}, \ldots, z_s)$ and $(z_1, \ldots, z_{i-1}, y^*, z_{i+1}, \ldots, z_s)$ are two points in S with $y = t - 1$ and $y^* = t$, then the points have the same color.

By construction,

$$(x_1, \ldots, x_{i-1}, y, z_{i+1}, \ldots, z_s) \text{ and } (x_1, \ldots, x_{i-1}, y^*, z_{i+1}, \ldots, z_s)$$

have the same color for *any* Shelah points $x_1 \in C_t^{N_1}, \ldots, x_{i-1} \in C_t^{N_{i-1}}$. But since each z_j, $j < i$, is on the Shelah line L^j, the z_j are clearly Shelah points, from which the claim follows immediately.

Therefore, if we set the N_i as required in the construction,

$$N_1 := r^{t^{s-1}},$$

(3.7) $\qquad M_i := t^{s-1}\prod_{j=1}^{i-1} N_j^2 \quad \text{and} \quad N_i := r^{M_i} \qquad (1 < i \leq s),$

$$N := \sum_{i=1}^{s} N_i,$$

the existence of the constructed Shelah space is ensured and the lemma is proved. □

The Hales-Jewett theorem can be deduced rather easily now.

Proof of the Hales-Jewett theorem. For any fixed r, we know that $HJ(1,r) = 1$. Now assume that $s := HJ(t-1, r)$ exists and let N be defined as in the statement of Lemma 3.38. Let c be an r-coloring of C_t^N. Lemma 3.38 guarantees the existence of a Shelah s-space $\Sigma_s = (L^1, \ldots, L^s)$ on which c is t-blind.

As before, we let $\pi : C_t^s \to \Sigma_s$ be the canonical bijection between C_t^s and Σ_s. We can pull back c to a coloring c^* on C_t^s by letting $c^* = \pi \circ c$. By choice of s and Lemma 3.36, C_t^s has a monochromatic line. The image of this line under π is a monochromatic line in Σ_s, which is in C_t^N. □

As indicated earlier, Shelah's proof gives a primitive recursive bound on $HJ(t,r)$ (and hence also on $W(k,r)$).

Corollary 3.39. *For every $r \geq 1$, $HJ(t,r)$ (as a function of t) is eventually dominated by $\Phi_5(t)$.*

Proof. Let $S(t)$ be the function defined in the proof of Lemma 3.38, that is, let $S(1) = 1$. For $s = S(t-1)$, let $S(t) = N$, where N is as in the last line of (3.7). Inspecting the definition of S, we see that from some point on, $S(t)$ is significantly larger than both r and t. In particular, with $s = S(t-1)$ we have

$$N_1 = r^{t^{s-1}} \leq s^{s^s} \leq \Phi_3(s)$$

and

$$M_i = t^{s-1} \prod_{j=1}^{i-1} N_j^2 \leq (N_{i-1})^{3s},$$

and thus, for sufficiently large t (and hence sufficiently large s),

$$N_i \leq r^{(N_{i-1})^{3s}} \leq 2^{2^{2^{N_{i-1}}}}.$$

Every iteration of N_i adds three more exponents to an exponential tower (the $\uparrow\uparrow$ function), and hence adds 3 to an argument of Φ_3. Therefore,

$$N_s \leq \Phi_3(s + 3(s-1)) \leq \Phi_3(4s),$$

and

$$S(t) = N = N_1 + \cdots + N_s \leq 2^{N_s} \leq \Phi_3(4s + 1).$$

A direct calculation shows that $S(3) \leq \Phi_4(8)$. So if we assume, inductively, that $S(t-1) \leq \Phi_4(2t)$, then

$$S(t) \leq \Phi_3(4s + 1) = \Phi_3(4S(t-1) + 1)$$
$$\leq \Phi_3(4\Phi_4(2t) + 1) \leq \Phi_3(\Phi_4(2t + 1)) = \Phi_4(2(t+1)).$$

3.5. A really fast-growing Ramsey function

Therefore $S(t)$ is eventually dominated by a function that is a combination of two functions from $\mathcal{E}_5$ and therefore, by Theorem 3.24, is eventually dominated by Φ_5. □

Of course, better bounds can be extracted from Shelah's proof, but here we were interested only in giving an explicit bound, as simply as possible, in terms of the Grzegorczyk hierarchy. By Proposition 3.34, the van der Waerden bound $W(k,r)$ is also in $\mathcal{E}_5$.

3.5. A really fast-growing Ramsey function

Using a seemingly small tweak to Ramsey's theorem, one can ensure that the corresponding Ramsey numbers grow faster than any primitive recursive function. This was shown by Paris and Harrington [47] and forms an integral part of their metamathematical analysis of Ramsey-type theorems, which we will describe in Chapter 4.

Let us call a set $Y \subseteq \mathbb{N}$ **relatively large** if

$$|Y| > \min Y.$$

Theorem 3.40 (Fast Ramsey theorem [47]). *For all integers $m, p, r \geq 1$, there exists $N \in \mathbb{N}$ such that whenever $[N]^p$ is r-colored, there exists a relatively large, homogeneous subset $Y \subseteq [N]$ of size at least m.*

The only difference from the usual finite Ramsey theorem (Theorem 1.31) is that the homogeneous set Y is required to be relatively large. Given $m, p,$ and r, we denote by $PH(m, p, r)$ the least N with the property asserted in the statement of the theorem.

Proof. The proof is a simple application of the compactness principle. We follow the blueprint given in Section 2.2.

Suppose that for some m, p, and r, there is no N which satisfies the statement of the theorem. For each natural number n, let

$T_n = \{f : [n]^p \to r$ such that there is

no homogeneous Y which is relatively large$\}$.

This set is finite for all n. Moreover, for each $f \in T_{n+1}$ there is a unique $g \in T_n$ such that f is an extension of g, that is, $g \subset f$. Therefore, we

have that the partially ordered set
$$T := \bigcup T_n$$
is a finitely branching tree which, by the assumption that T_n is non-empty for all n, is infinite.

By König's lemma (Theorem 2.6), we have an infinite path
$$f_1 \subset f_2 \subset \cdots$$
in T with each $f_i \in T_i$. Let $f = \bigcup f_i$, which is an r-coloring of the natural numbers. By the infinite Ramsey theorem (Theorem 2.1), there exists an infinite, homogeneous $X \subset \mathbb{N}$, say
$$\{x_1 < x_2 < x_3 < \cdots\}.$$
If we now restrict this set to the first $s := x_1 + 1$ elements
$$\{x_1, x_2, \ldots, x_s\},$$
the set is a homogeneous subset of $[x_s]$ which is relatively large. This is a contradiction. $\square$

We will show that PH grows very fast. To facilitate the analysis, we introduce a slight variant of the function family $\{\Phi_n : n \in \mathbb{N}\}$. Define functions Ψ_n by
$$\Psi_0(x) = x + 1,$$
$$\Psi_{n+1}(x) = \Psi_n^{(x)}(x).$$

Compare this with Definition 3.23.

Exercise 3.41. Show that $\Psi_1(x) = 2x$ and $\Psi_2(x) = x \cdot 2^x$.

The functions Ψ_n have a growth behavior similar to the Φ_n. In fact, each Ψ_n appears at the same level of the Grzegorczyk hierarchy $\{\mathcal{E}_n : n \in \mathbb{N}\}$ as Φ_n.

Theorem 3.42. For every $n \geq 1$, $PH(n+2, 2, n+1) \geq \Psi_n(n+1)$.

Proof. We define a 2-coloring c_1 as follows: Split the set $[2, \infty)$ into intervals $[x, 2x)$, i.e.
$$[2,4) \cup [4,8) \cup [8,16) \cup \cdots.$$

3.5. A really fast-growing Ramsey function

Call these intervals the *type 1* blocks. If i and j lie in the same type 1 subset, put $c(i,j) = 1$. Otherwise, let $c_1(x,y) = 0$.

If A is a homogeneous subset of color 1, then $A \subset [x, 2x)$ for some x, which means that $|A| \leq x$ and hence A is not relatively large. Therefore, any relatively large, homogeneous subset A must have color 0, and if $|A| \geq 3$, then A has to contain at most one element each from $[2,4)$ and $[4,8)$. Therefore, $PH(3,2,2) \geq 4 = \Psi_1(2)$.

Next, we define a type 2 block structure in a similar way. Split $[2, \infty)$ into sets $[x, x \cdot 2^x)$ (note that $\Psi_2(x) = x \cdot 2^x$):

$$[2, 8) \cup [8, 8 \cdot 2^8) \cup \cdots.$$

We keep the c_1 color if $c_1(x, y) = 1$. Additionally, we color the set $\{i, j\}$ with color 2 if i and j are in the same type 2 block but not the same type 1 block.

Suppose A is a relatively large, homogeneous set. We argue as before that the color of A cannot be 1. If it is 2, A must be a subset of some interval $[x, x \cdot 2^x)$. This interval is split into type 1 subintervals,

$$[x, 2x) \cup [2x, 4x) \cup \cdots \cup [x \cdot 2^{x-1}, x \cdot 2^x).$$

As we see, there are $(x-1)$ such subintervals, and A contains at most one element from each of these type 1 blocks (otherwise there would be a pair of color 1). Therefore, $|A| \leq x$ holds in this case, too. It follows that any relatively large, homogeneous subset of cardinality greater than or equal to 4 must have at most one element from each of

$$[2, \Psi_2(2)), [\Psi_2(2), \Psi_2(\Psi_2(2))), [\Psi_2^{(2)}(2), \Psi_2^{(3)}(2)),$$

and therefore

$$PH(4, 2, 3) \geq \Psi_2^{(2)}(2) \geq \Psi_2(3).$$

We can now continue inductively. The type $n + 1$ blocks have the form $[x, \Psi_{n+1}(x))$ (where x itself is of the form $\Psi^{(k)}(3)$), and to define coloring c_{n+1} we keep c_n except that we color $\{i, j\}$ with color $n + 1$ if both i and j lie in the same type $s + 1$ block but not the same type s block.

By the definition of Ψ_{n+1},

$$[x, \Psi_{n+1}(x)) = [x, \Psi_n^{(x+1)}(x))$$
$$= \underbrace{[x, \Psi_n(x)) \cup \cdots \cup [\Psi_n^{(x)}(x), \Psi_n^{(x+1)}(x))}_{(x-1) \text{ intervals}}.$$

An argument similar to the one in the $n = 2$ case then yields

$$PH(n+3, 2, n+2) \geq \Psi_{n+1}(n+2),$$

which, by way of induction, proves the theorem. □

Corollary 3.43. *The function* $f(x) = PH(x+3, 2, x+1)$ *eventually dominates* Ψ_n, *for all* n, *and therefore also every primitive recursive function.*

One can continue along these lines for higher values of p (that is, look at p-sets instead of just pairs). The analysis becomes much more involved. In a landmark paper, Ketonen and Solovay [40] were able to show that

$$PH(x+1, 3, x) \text{ eventually dominates every } \Phi_\alpha(x) \text{ with } \alpha < \omega^\omega,$$
$$PH(x+1, 4, x) \text{ eventually dominates every } \Phi_\alpha(x) \text{ with } \alpha < \omega^{\omega^\omega},$$
$$\vdots$$

The diagonal function $PH(x+1, x, x)$ will then eventually dominate every function $\Phi_\alpha(x)$ with $\alpha < \varepsilon_0$.

Ketonen and Solovay's argument is an elementary combinatorial analysis, but rather complicated and long, and we will not reproduce it here. We hope nevertheless to have conveyed some of the explosive growth that is produced by requiring homogeneous sets to be relatively large. The basic idea is that certain colorings can force a relatively large homogeneous set to have rather large gaps. This is a phenomenon we will encounter again.

On its own, the results of this section might appear to be a mere mathematical curiosity—a slight variation on Ramsey's theorem causing explosive growth of the accordant Ramsey numbers. Their significance lies in the ties to the metamathematics of arithmetic, and

3.5. A really fast-growing Ramsey function

to the seminal developments in this area in the 20th century, most notably Gödel's incompleteness theorems. It takes a bit of work to develop the necessary background for this, and we will embark on it in the next chapter.

Chapter 4

Metamathematics

4.1. Proof and truth

When do we consider a mathematical proposition to be true?
The longer one thinks about it, the more delicate this question may become.

Some statements seem to be so obvious that their truth is self-evident, such as the fact that there is no greatest natural number. For other statements, less obvious ones, we require a proof to convince ourselves. As an example, take the proposition that there is no greatest prime number.

But what if we cannot find a proof, nor are we able to disprove the assertion? The famous open mathematical problems come to mind. Consider the *Goldbach conjecture*, for example, which states that every even integer greater than 2 can be written as the sum of two primes. This is a perfectly reasonable mathematical statement, and probably most mathematicians would say that the conjecture either holds or fails, but we just do not know yet which is the case. When light is finally shed on the Goldbach conjecture, it will be in the form of a *proof* of either the statement

$$\forall m \in \mathbb{Z} \, (m > 2 \text{ and } m \text{ even} \Rightarrow \exists p_1, p_2 \text{ prime } (p_1 + p_2 = m))$$

or its negation

$$\exists m \in \mathbb{Z} \, (m > 2 \text{ and } m \text{ even and } \forall p_1, p_2 \text{ prime } (p_1 + p_2 \neq m)).$$

Note that to prove the negation, all we have to do is exhibit a single even number m and check for all pairs of prime numbers smaller than m that they do not sum to m. It is a simple task that a computer could do. And mathematicians have been using computers to search for such an m, but so far, they have not found one.[1]

But why should we assume that a proof will in fact be found? Could it even be that *no proof exists*? What would that even mean?

It is a remarkable achievement of mathematical logic to not only rigorously define what constitutes a mathematical proof (in other words, proofs are established as mathematical objects, the same way that continuous functions are mathematical objects) but also show that there are statements which, in a certain well-defined sense, *cannot be settled by proof at all*, that is, neither the statement nor its negation is provable.

What exactly constitutes a proof? We have already given plenty of proofs in this book, and if we try to extract the essential features common to all of them, we might come up with the following list:

- A proof derives the statement we want to prove from other statements, either statements we have proved before, or basic assumptions we make about the objects we are dealing with (usually referred to as "axioms").
- The derivation consists of *finitely many steps*. In each step we advance the argument by applying a *logical rule*, such as if A implies B and A holds, then B holds.
- Each step should be so simple that everyone, even a computer, can check its validity.
- If we prove something, it should of course be *true*.

While the first point is hopefully acceptable, the other three are more problematic. Are proofs really finite? Of course, no one would be able to read a proof of infinite length, but don't we sometimes have to check infinitely many cases? We do not check them all in the end, but usually do one case and then argue something like "the other cases are similar". Furthermore, not every step in a proof seems to

---

[1] At the time this book was written, all numbers up to 4×10^{17} have been checked.

4.1. Proof and truth

be the application of a simple logical rule. When we take derivatives, we might apply (sometimes highly sophisticated) symbolic rules. And finally, as you may have experienced yourself, proofs given in books or papers are often not as simple and clear as one would wish. Often details, important or not, are left out. While we might blame the authors for that, one could reply that many mathematical arguments are so complex that spelling them out completely would require hundreds or even thousands of pages, most of which are not necessary for a mathematically educated reader. But the idea is that we *could do just that for every proof.* After all, as Weinberger [**72**] writes, "proofs should not be a matter of opinion". But this still leaves the question of when *are* the steps simple enough for everyone, even a computer?

It turns out we can avoid these pitfalls if we regard proofs as a *completely syntactical* affair. If we can regard every mathematical statement featured in the stages of a proof as a sequence of symbols, and we agree on a fixed set of logical rules we can use to pass from one stage to another, then a step in the proof (i.e. the application of a rule) amounts to a *manipulation of symbols*. And this can indeed be checked algorithmically.

Such an approach, however, requires that we can actually completely formalize mathematical statements. We give a brief introduction on how to do this in the next section.

But before we progress, we want to point out that, while formalization may enable us to rigorously specify what a mathematical proof is, it might result in the loss of the desired connection between proof and truth—purely symbolic relations need not have any semantic content. A formula itself is just a sequence of symbols. It has no meaning, just as the word "lamp" is just a sequence of letters, 'l', 'a', 'm', and 'p'. We attach meaning to a word by relating it to objects in the real world, in this case the collection of all lamps. Similarly, we give meaning to a mathematical formula by interpreting it in the "real mathematical world", associating it with a specific mathematical object. This way we can say that a formula is *true* relative to this object.

We will make all this more precise below; here we just want to indicate that once a formal definition of proof is given, the connection between proof and truth needs to be rigorously (re-)established. That all this is possible is one of the great triumphs of mathematical logic in the 20th century.

Mathematical syntax. Formal mathematical statements are formed according to *rules*. If you have learned a programming language, this concept will be familiar, as the rules for forming mathematical statements are very similar to the rules that govern the formation of a valid Python program, for example.

It is important that we agree upon a *fixed set of symbols* that we are allowed to use.

The basic symbols provided by **first-order predicate logic** are always available:

- logical symbols: $\forall$ (for all), $\exists$ (exists), $\wedge$ (and), $\vee$ (or), $\neg$ (not), $\Rightarrow$ (implies), (,), =;
- variable symbols: $x_0, x_1, x_2, \ldots$.

The **non-logical symbols** will depend on what kind of mathematical objects we want to study. Suppose we are interested in groups. What kind of symbols do we need to express statements about groups? Every group has a binary operation that takes two elements of a group and assigns them a third element of the group (possibly equal to one of the two). For example, in the additive group of the integers $\mathbb{Z}$, this operation is addition, while for the general linear group $\mathrm{GL}(n, \mathbb{R})$, that operation is matrix multiplication. Furthermore, every group has a neutral element with respect to the group operation. This suggests that we need two symbols to talk about groups, one for the group operation, say '$\circ$', and one for the neutral element, say 'e'. The *language of group theory* has the non-logical symbols

$$\circ, \quad e.$$

In this book we are interested in the natural numbers, and we will consider three basic operations:

- the *successor operation* $S : x \mapsto S(x) = x + 1$,

4.1. Proof and truth

- addition $(x, y) \mapsto x + y$,
- multiplication $(x, y) \mapsto x \cdot y$.

Furthermore, the number 0 (which we count among the natural numbers) has a special status, because, for example:

- it is the smallest natural number;
- it is the only natural number that is not the successor of another natural number;
- adding it to another number does not change that number.

Therefore, the **language of arithmetic** $\mathcal{L}_A$ has the four symbols

$$S, \quad +, \quad \cdot, \quad 0.$$

For other mathematical structures, we would choose yet a different set of non-logical symbols. For orders, we would need '<'; for set theory, '∈'.

Formulas are built from symbols via rules. What these rules are specifically, and how they are used to form formulas, is not important at this point (although we have to consider it in more detail later on). All that matters for now is that the rules enable us to distinguish between valid formulas, such as

$$\forall x_0 \forall x_1 (x_0 + x_1 = x_1 + x_0), \quad \exists x_0 (x_0 = x_0), \quad 0 + x_5 = x_5,$$

and invalid ones, such as

$$()x_6 = 00), \quad (((((())))))), \quad x_0 x_7 ==.$$

Furthermore, we can write a computer program to check whether a finite sequence of symbols from our symbol set is a valid formula. (This is exactly the task done by a syntax checker when you compile a computer program.)

We have to address the relation between variables and quantifiers. Variables take values in the mathematical structure we are analyzing. The truth of a formula may depend on the value of the variable. For example, in $\mathbb{N}$,

$$x_0 + 3 = 4$$

is true for $x_0 = 1$ but false for $x_0 = 2$. Quantifying a variable removes this uncertainty.

$$\exists x_0 \, (x_0 + 3 = 4)$$

represents the statement "for **some** x_0, the equation $x_0 + 3 = 4$ holds". There is no more ambiguity—the statement is either true or false (true, in this case). Similarly,

$$\forall x_0 \, (x_0 + 3 = 4)$$

means "for **all** x_0, the equation $x_0 + 3 = 4$ holds" (which is false). In both examples, we say that the variable x_0 is **bound** by the quantifier $\exists$ or $\forall$, respectively.

Accordingly, a variable is called **free** if it is not bound by a quantifier. In $x_0 + 3 = 4$, the variable x_0 occurs free. A formula can have both free and bound variables. In the formula

$$\exists x_0 \, (x_1 + x_0 = 0),$$

the variable x_1 is free while the variable x_0 is not. A **sentence** is a formula with no free variables. In the three examples of valid formulas above, the first two are sentences while the third is not. Formulas like

$$0 + 0 = 0$$

are also sentences because they have no variables at all.

Remark. Strictly speaking, the notions "formula" and "sentence" make sense only if a language $\mathcal{L}$ is given. Therefore, we should always be speaking of "$\mathcal{L}$-formulas" and "$\mathcal{L}$-sentences". However, in the following we will often suppress the "$\mathcal{L}$-", because the language is either clear or irrelevant.

Axiom systems. An *axiom system* is simply a set of sentences in a fixed language $\mathcal{L}$. An axiom system may be finite or infinite, but in any case we require that we be able to recognize algorithmically whether an $\mathcal{L}$-formula is an axiom. Similar to the syntax checker, which checks whether a finite sequence of symbols is a valid $\mathcal{L}$-formula, there should be an *axiom checker*, a computer program that decides whether a given $\mathcal{L}$-formula is an axiom.

4.1. Proof and truth

If the axiom system is finite, this will be no problem. For example, take the language of group theory, $\mathcal{L}_G = \{\circ, e\}$. The *group axioms* can be formalized as

(G1) $\forall x_0 \, (x_0 \circ e = x_0 \wedge e \circ x_0 = x_0)$,
(G2) $\forall x_0, x_1, x_2 \, (x_0 \circ (x_1 \circ x_2) = (x_0 \circ x_1) \circ x_2)$,
(G3) $\forall x_0 \exists x_1 \, (x_0 \circ x_1 = e \wedge x_1 \circ x_0 = e)$.

We can simply store all three axioms in the memory of the machine and compare any given formula successively with each of them. However, this procedure is impossible if the axiom system is infinite. In this case the sentences of the system must be describable in a systematic (algorithmic) way.

The axiom system we will be particularly interested in is **Peano arithmetic** (PA), formalized in the language of arithmetic $\mathcal{L}_A$ and described by the following axioms:

(PA1) $\forall x \, S(x) \neq 0$
(PA2) $\forall x \, \forall y \, (S(x) = S(y) \Rightarrow x = y)$
(PA3) $\forall x \, (x + 0 = x)$
(PA4) $\forall x \, \forall y \, (x + S(y) = S(x + y))$
(PA5) $\forall x \, (x \cdot 0 = 0)$
(PA6) $\forall x \, \forall y \, (x \cdot S(y) = x \cdot y + x)$

In addition to (PA1)–(PA6), the system PA also includes an *induction scheme*: For *every $\mathcal{L}_A$-formula $\varphi(x, \vec{y})$* (which may have any number of free variables $\vec{y} = y_1, \ldots, y_n$), the sentence

(Ind_φ) $\quad \forall \vec{y} \left[\left(\varphi(0, \vec{y}) \wedge \forall x \, (\varphi(x, \vec{y}) \to \varphi(S(x), \vec{y})) \right) \to \forall x \, \varphi(x, \vec{y}) \right]$

is an axiom of PA. As there are infinitely many $\mathcal{L}_A$-formulas, this results in infinitely many different axioms (Ind_φ). We are therefore not able to list them all here (after all this book has only finitely many pages), but the mechanism by which they are added is clear. This is why it is called a *scheme*. In particular, an axiom checker for PA, as required above, exists: Given an $\mathcal{L}_A$-formula ψ, we first compare ψ (symbol by symbol) with any of the six axioms (PA1)–(PA6). If ψ is not among those, we parse ψ to see if it is of the form prescribed by the scheme (Ind), identifying a subformula φ with $\psi \equiv \text{Ind}_\varphi$.

The axioms (PA1)–(PA6) describe some basic properties of natural numbers. In particular, they ensure that addition is defined recursively from the successor operation and multiplication is defined recursively from addition, as we already observed in Section 3.2. The scheme (Ind) ensures that this method of induction actually works for any property or function defined by a formula.

Induction comes in many guises, and towards the end of this chapter we will use a principle equivalent to induction, the **least number principle**:

$$(\text{LNP}_\varphi) \quad \forall \vec{w}[\exists v\, \varphi(v, \vec{w}) \rightarrow \exists z(\varphi(z, \vec{w}) \wedge \forall y < z\, \neg \varphi(y, \vec{w}))]$$

In other words, the LNP says if a formula has a witness, it has a *least* such witness. It is not hard to find structures in which the least number principle fails. Take for example the real numbers $\mathbb{R}$ and let $\varphi(v)$ be the statement $v > 0$. However, the LNP is equivalent to induction. One can show this formally by introducing a new kind of axiom system in which essentially the induction scheme (Ind_φ) is replaced by the scheme (LNP_φ) and then showing that this axiom system proves all instances of (Ind_φ), and vice versa.

Exercise 4.1. Argue (informally) that induction and the least number principle are equivalent.

Remark: The careful reader may already have noticed that we have become a little sloppy concerning formal notation—we left out a parenthesis here and there, and we used x, y to denote variables instead of x_0 and x_1. All this is done to improve readability.

Proofs. Suppose that A is an axiom system for some fixed language $\mathcal{L}$ and σ is an $\mathcal{L}$-sentence. We want to define what a *proof* of σ from A is.

Definition 4.2. A **proof of** σ **from** A is a sequence $(\varphi_1, \ldots, \varphi_n)$ such that $\varphi_n = \sigma$ and for all $i < n$, φ_i either

4.1. Proof and truth

(1) is an axiom $\psi \in A$, or

(2) is a *logical axiom*, i.e. a formula from a fixed set of universally valid formulas such as $x = x$ or $\psi \vee \neg \psi$, or

(3) has been obtained from formulas $\varphi_1, \ldots, \varphi_{i-1}$ by application of a *deduction rule*. For example, if φ_2 is $\varphi_1 \Rightarrow \psi$, then we are allowed to deduce ψ. In other words, if we have previously established φ_1 and $\varphi_1 \Rightarrow \psi$, we may deduce ψ by applying the logical rule called *modus ponens*.

The choice of the logical axioms (2) and the choice of deduction rules (3) define a **proof system**. The precise nature of such a proof system is of no importance here. What matters for us is that the system is *sound* and *complete* (properties we will discuss below). Readers can find various such systems in the books by Shoenfield [**60**] and Rautenberg [**54**].

In mathematical practice, a proof is usually not given in this form. It would be very hard to digest for a human reader who wants to follow an argument. Proofs in mathematical papers and books (such as this one) are usually given in a hybrid form: formal computations paired with a deduction given in English or another language. But the idea is that every proof *can* be brought into the form of Definition 4.2. And once this is done, it should not be very hard (though it would be tedious) to go through the proof literally line by line to check whether every step is valid. In fact, we could leave this task to a computer.

In recent years enormous progress has been made in developing *proof assistants*[2], which help humans to turn their "human" proofs into fully formal arguments the correctness of which can then be checked by computers. Several important theorems of mathematics have successfully been verified this way, for example proofs of the Kepler conjecture in geometry [**30**] and the Feit-Thompson theorem in group theory [**21**].

Truth. If a mathematical statement is just a sequence of symbols and a mathematical proof is just a sequence of statements—that is,

---

[2] Coq, HOL, Isabelle, Lean, to name just a few.

a sequence of sequences of symbols, hence itself just a sequence of symbols—how do we guarantee that a statement we proved is *true*?

A sentence itself is neither true nor false (it is just a sequence of symbols, strictly speaking). It is given meaning by interpreting it in a mathematical structure. Let us consider the sentence

$$\forall x_0 \exists x_1 (x_0 + x_1 = 0).$$

This sentence is false when interpreted in the natural numbers $\mathbb{N}$, since adding two natural numbers at least one of which is positive will always result in a positive number. On the other hand, if we interpret the sentence in the integers $\mathbb{Z}$, it would be true since we can simply choose $x_1 = -x_0$.

In general, to give meaning to a sentence in this way requires two things:

(i) We need to specify a set, called the *universe*, over which our variables range.

(ii) We need to *interpret the non-logical symbols*. For example, we need to specify what the symbol '+' means in our universe.

Usually, the choice of symbols indicates what interpretation we have in mind (such as '+' meaning addition in $\mathbb{N}$ and $\mathbb{Z}$). But in other cases, the interpretations can be quite different. For example, if we are studying groups, we have the symbol ∘, and we can interpret it as addition over the real numbers, but we could also interpret it as matrix multiplication for the group of all invertible $n \times n$ matrices.

Given a language $\mathcal{L}$, a universe together with an interpretation of the symbols is called an $\mathcal{L}$-**structure**.

If a sentence σ holds for a given structure $\mathcal{A}$, we write

$$\mathcal{A} \vDash \sigma$$

and say that "$\mathcal{A}$ models σ" or "σ is true in $\mathcal{A}$". For example,

$$\mathbb{Z} \vDash \forall x_0 \exists x_1 (x_0 + x_1 = 0) \quad \text{but} \quad \mathbb{N} \nvDash \forall x_0 \exists x_1 (x_0 + x_1 = 0).$$

Here, you have to think of $\mathbb{Z}$ and $\mathbb{N}$ not only as sets, but as sets together with the operation +.

4.1. Proof and truth

More generally, if $\varphi(x_1,\ldots,x_n)$ is an $\mathcal{L}$-formula with free variables $x_1,\ldots,x_n$, $\mathcal{A}$ is an $\mathcal{L}$-structure, and $a_1,\ldots,a_n \in A$, we write

$$\mathcal{A} \vDash \varphi[a_1,\ldots,a_n]$$

to indicate that φ holds in $\mathcal{A}$ if the free variables $x_1,\ldots,x_n$ are evaluated as $a_1,\ldots,a_n$, respectively.

For example, let $\varphi(x_1, x_2) \equiv \exists y\,(x_1 + x_2 = 2y)$.[3] Then

$$\mathbb{Z} \vDash \varphi[2,4] \quad \text{but} \quad \mathbb{Z} \nvDash \varphi[2,5].$$

Models of Peano arithmetic. A **model** of an axiom system S is a mathematical structure $\mathcal{A}$ in which every sentence of the axiom system is true. In other words,

$$\text{for all } \sigma \in S,\ \mathcal{A} \vDash \sigma.$$

For example, a model of the group axioms (G1)–(G3) is simply a mathematical group.

We are interested in models of Peano arithmetic, so let us think about what they might look like.

The Peano axioms are intended to capture arithmetic operations on the natural numbers, and obviously the structure $\mathbb{N}$, in which we interpret the symbol '+' as addition on $\mathbb{N}$, '·' as multiplication on $\mathbb{N}$, 'S' as the successor function $S : x \mapsto x + 1$, and '0' as the natural number 0, satisfies all the axioms (PA1)–(PA6). $\mathbb{N}$ also satisfies all induction axioms (Ind$_\varphi$): Suppose that for some formula $\varphi(x, \vec{y})$, the conclusion $\forall x \varphi(x, \vec{y})$ is violated. Then there exists a least x for which this happens, and this in turn yields that either $\varphi(0, \vec{y})$ does not hold or $\forall x\,(\varphi(x, \vec{y}) \to \varphi(x+1, \vec{y}))$ is violated.

The **standard model** of arithmetic is the set (universe) $\mathbb{N}$ of natural numbers with the successor function $S(n) = n+1$, addition $n+m$, multiplication $n \cdot m$, and '0' interpreted as the natural number 0.

---

[3] You may have noticed the use of ≡ here to denote equality *between* formulas. This is to distinguish it from the logical symbol =, which is used *inside* formulas.

Our intention is to prove facts about the natural numbers while only assuming very basic facts about them. So, do we really *need* the induction axioms?

It turns out that if we drop the induction scheme, we can have models that look a bit like the natural numbers, but which can still be very different. Consider, for example, the set $\mathbb{R}^{\geq 0}$, consisting of all non-negative real numbers.

Exercise 4.3. Verify that with the usual operations $x \mapsto x+1, +, \cdot, 0$, $\mathbb{R}^{\geq 0}$ forms a model of (PA1)–(PA6).

But clearly $\mathbb{R}^{\geq 0}$ is in many respects very different from $\mathbb{N}$. For example, consider the statement

(4.1) $$\forall x (x \neq 0 \to \exists y\, S(y) = x),$$

which says that if a number is not zero, it is the successor of another number. This is clearly true for the standard model of arithmetic, if we interpret $S(y)$ as $y+1$ (and it can be proved using *induction*). But it does not hold for $\mathbb{R}^{\geq 0}$: $1/2$ is not a successor of anything, since we are only allowing *non-negative* real numbers.

Now you may suggest: (4.1) *is a simple enough statement, and clearly true of* $\mathbb{N}$*. Why do we not add it as axiom* (PA7) *to our list?*

While this certainly makes sense, it will not rule out "undesired" models completely.

$\mathbb{Z}[X]$ is the set of all polynomials with integer coefficients. Let $\mathbb{Z}[X]^+$ be the subset of $\mathbb{Z}[X]$ that includes the zero polynomial $p(x) \equiv 0$ and all polynomials the leading terms of which have positive coefficients, i.e. if

$$p(x) = a_n x^n + a_{n-1} x^{n-1} + \cdots + a_0$$

(and $p(x)$ is not the zero polynomial), we require that $a_n > 0$. How would we interpret $S, +, \cdot, 0$ over this set?

Now the symbol '0' must be a *polynomial*, and it is obvious to take the zero polynomial. We can add and multiply polynomials, too, so let us interpret $+$ and $\cdot$ as polynomial addition and multiplication, respectively. But what should be the "successor" of a polynomial? Let us try the following: Put $S(p) := p + 1$, i.e.

$$S(a_n x^n + a_{n-1} x^{n-1} + \cdots + a_0) = a_n x^n + a_{n-1} x^{n-1} + \cdots + (a_0 + 1).$$

4.1. Proof and truth

Exercise 4.4. Show that $\mathbb{Z}[X]^+$ with these interpretations satisfies (PA1)–(PA6) and (4.1) a.k.a. (PA7).

(*Hint:* Polynomials in $\mathbb{Z}[X]^+$ may have negative coefficients; just the leading one has to be positive.)

Exercise 4.5. Find other models, different from the standard model, which satisfy as many axioms from (PA1)–(PA6) as possible. Try adding additional elements to the standard model "at the end" and extend the operations $S, +, \cdot$ accordingly.

If we add the induction scheme (Ind_φ), $\mathbb{Z}[X]^+$ ceases to be a model.

Exercise 4.6. (a) Use induction to show that $\mathbb{N} \vDash \forall x \exists y\, (2y = x \lor 2y + 1 = x)$. In other words, every natural number is either even or odd. (It is not hard, albeit a little tedious, to turn this into a a formal proof from **PA**.)

(b) Show that the above formula does not hold in $\mathbb{Z}[X]^+$. Conclude that $\mathbb{Z}[X]^+$ is not a model of **PA**.

(*Hint:* Consider the polynomial $p(x) = x$.)

Does the axiom system **PA** have any models other than the standard model $\mathbb{N}$?

The answer is, maybe a bit surprisingly, "yes". To see why, we need to return to our brief introduction to mathematical logic and talk about the relation between proof and truth.

Proof versus truth. If a statement σ is provable from an axiom system A, is the statement true? Now that we have a mathematical definition of truth, we can make this question precise.

Let us write
$$A \vdash \sigma$$
to denote that σ is provable from A. On the other hand, we introduced the notation
$$A \vDash \sigma$$
to express that

σ holds in every structure that satisfies the axioms of A.

For example, if G is the set of group axioms (G1)–(G3), then $G \vDash \sigma$ means σ is a statement that *holds in every group*. If this is the case, we can say that σ is *a logical consequence of the group axioms*. Furthermore, σ is what is usually called a theorem of group theory. For this reason, axiom systems are also called **theories**.

We can now formulate the requirement that whatever we prove is true as

$$A \vdash \sigma \text{ implies } A \vDash \sigma.$$

If this is the case, we call our proof system **sound**. It is usually not hard to establish the soundness of a proof system. The reason is that the transformations allowed in a proof are of a logically simple nature.

But if we consider establishing logical consequences as our main goal (such as establishing theorems about groups), then soundness is the bare minimum we should expect of our formal notion of proof. What we really want is the opposite direction: *if something is a logical consequence of the axioms, there should be a proof for it.* More formally, for any axiom system A and any sentence σ in the same language,

$$A \vDash \sigma \text{ implies } A \vdash \sigma.$$

This property is referred to as the **completeness of the proof system**. We have to be very careful with this notion, as there is also the property of *completeness of a theory*, which will play an important role later.

It was one of Kurt Gödel's many remarkable contributions to mathematical logic to show that there exist proof systems that are sound and complete.

Theorem 4.7 (Gödel completeness theorem). *For any axiom system A,*

$$A \vdash \sigma \text{ if and only if } A \vDash \sigma.$$

A proof of the completeness theorem can be found in numerous textbooks on logic (for example, [**13**, **54**, **60**]).

The completeness theorem is a truly remarkable fact. Consider the task of establishing a theorem about groups, i.e. a logical consequence of the group axioms. Following the definition of logical

4.1. Proof and truth

consequence, this would mean checking for every group whether the statement holds in that particular group. But there are way too many groups. In fact, the family of all mathematical groups is not even a set, but a proper class, just like the class of all ordinals.

Of course, nobody proves theorems about groups this way. We deduce them from the group axioms. What the completeness theorem tells us is that

> *every statement true in all groups has a proof from the group axioms, and this proof can be completely formalized.*

The completeness theorem also has some important consequences at the other extreme: inconsistent theories.

A theory T is **inconsistent** if for some sentence σ,
$$T \vdash \sigma \text{ and } T \vdash \neg\sigma.$$
(Note that, by the completeness theorem, we could use $\vDash$ instead of $\vdash$.) In any fixed mathematical structure $\mathcal{M}$, we have *either* $\mathcal{M} \vDash \sigma$ *or* $\mathcal{M} \vDash \neg\sigma$, but never both (since a sentence is either true or not, in which case its negation is true). Therefore, if T is inconsistent, it cannot have any models. The other direction of the completeness theorem tells us in turn that if a theory does not have any models, then it must be inconsistent.

Corollary 4.8. *A theory T is consistent if and only if T has a model.*

Exercise 4.9. Show that if T is inconsistent, $T \vdash \sigma$ for every sentence σ. In other words, an inconsistent theory proves everything.

(*Hint:* If $T \not\vdash \tau$ for some τ, then there has to be some structure witnessing this.)

If T has a model, it is called **satisfiable**. The relation between consistency and satisfiability also hints at a possible way to show that a statement is *not* provable from a theory.

Lemma 4.10. *T does not prove σ if and only if $T \cup \{\neg\sigma\}$ has a model.*

Proof. ($\Rightarrow$) If $T \cup \{\neg\sigma\}$ has no model, it is inconsistent. This means that $T \cup \{\neg\sigma\}$ proves everything (see Exercise 4.9); in particular $T \cup \{\neg\sigma\} \vdash \sigma$. The *deduction theorem*, which can be proved without using the completeness theorem (see e.g. [60]), states that if T is a theory and τ and σ are sentences, then

$$T \cup \{\tau\} \vdash \sigma \text{ implies } T \vdash \neg \tau \vee \sigma.$$

Applying the deduction theorem to $\tau = \neg\sigma$, we see that $T \vDash \neg\neg\sigma \vee \sigma$, and since $\neg\neg\sigma$ is logically equivalent to σ, $T \vdash \sigma$, as desired.

($\Leftarrow$) If $T \cup \{\neg\sigma\}$ has a model, then there exists a model of T in which σ does not hold. Hence $T \nvDash \sigma$, and therefore by the completeness theorem, T does not prove σ. $\square$

Exercise 4.11. Show that the deduction theorem can be easily deduced from the completeness theorem.

To show that a statement is *not* provable from a set of axioms, it therefore suffices to find a model of the axioms in which the statement does not hold. But how do we find such models? At this point, a principle surfaces again that has played an important role throughout this book: *compactness*.

Compactness in first-order logic. The completeness theorem tells us that if a statement σ is a logical consequence of an axiom system A, then there is a formal proof of this. Since a proof has only finitely many steps, it follows that a proof can use at most finitely many axioms from A.

Corollary 4.12. *If $A \vDash \sigma$, then $A_0 \vDash \sigma$ for some finite $A_0 \subseteq A$.*

This is an easy observation, but it has an important consequence.

Theorem 4.13 (Compactness theorem of first-order logic). *Let T be a set of sentences in a language $\mathcal{L}$. If every finite subset of T has a model, then T has a model.*

Exercise 4.14. Deduce the compactness theorem from Corollary 4.12.

Note that the statement of the compactness theorem does not mention logical consequence (or, equivalently, proof) anymore. You may wonder why it is called the *compactness theorem*. As we saw in Chapter 2, compactness is a general finiteness principle, and the compactness theorem for logic says that the existence of a model for an infinite axiom system can be reduced to models for finite subsystems. In a more topological language (albeit loose), whenever T "covers" (i.e. implies) a sentence σ, there exists a finite subcovering T_0 of σ. One can indeed prove the compactness theorem as a purely topological result, and you can find a presentation of this in [57].

4.2. Non-standard models of Peano arithmetic

The compactness theorem has an important consequence for models of PA. In a nutshell, even by adding the induction axioms we cannot rule out the existence of strange models that look very different from the standard model $\mathbb{N}$. Such models are called, not surprisingly, **non-standard**, and they will play a crucial role in this chapter.

In the language of arithmetic $\mathcal{L}_A$, we have the constant symbol '0'. This means that we can directly name the special zero element in any $\mathcal{L}_A$-structure, and we can use it in formulas. But we can also name its successor using the term $S(0)$. In general, we can name the "number" n using the term $S(S(\ldots S(0)\ldots))$, by n applications of S. Let us write $\underline{n}$ as an abbreviation for this term. Note that $\underline{n}$ is just a formal expression, and depending on what $\mathcal{L}_A$-structure we are considering, it is not necessarily a number. For example, if we consider the structure $\mathbb{Z}[X]^+$, $\underline{n}$ will be the constant polynomial $p(x) \equiv n$. But, of course, in the standard model $\underline{n}$ denotes the natural number n. And since every model of PA must have a value for each term $\underline{n}$,

every model of PA contains a "version" of $\mathbb{N}$.

More precisely, if $\mathcal{N}$ is a model of PA, let

$$\mathbb{N}^{\mathcal{N}} = \{\underline{n}^{\mathcal{N}} : n \in \mathbb{N}\},$$

where $\underline{n}^{\mathcal{N}}$ denotes the interpretation of the expression $\underline{n}$ in the structure $\mathcal{N}$. Furthermore, since every model of PA satisfies axiom (PA4), we have that for any x,

$$x + \underline{1} = x + S(0) = S(x);$$

that is, in any model of PA the successor operation must be the same as addition by 1.

Now consider the sentence

(4.2) $$\varphi_n \equiv \exists x\, (x > \underline{n}).$$

The careful reader will notice that this is not an $\mathcal{L}_A$-formula, since in $\mathcal{L}_A$ we do not have a '<' symbol. This is not really a problem since we can simply define the usual order on $\mathbb{N}$ using + and 0:

(4.3) $$x < y :\Leftrightarrow \exists z\, (z \neq 0 \wedge x + z = y).$$

From now on we will use '<' just as it were another symbol of the language. But strictly speaking, we would have to replace every instance of it by its definition given above.

It is obvious that φ_n holds in the standard model, for every n. Let us reformulate this rather trivial fact a little bit, using the framework of languages, axioms, and their structures and models. We add a new symbol to the language of arithmetic, say 'c'. Let us call the new language $\mathcal{L}_A^c$,

$$\mathcal{L}_A^c = \{S, +, \cdot, 0, c\}.$$

c is a *constant symbol*, just like 0, meaning it will have to be interpreted as a *fixed element* in a structure of the new language.

We can use the new constant symbol in formulas, such as

$$\varphi_n^c \equiv c > \underline{n}.$$

We obtain an $\mathcal{L}_A^c$-structure by taking any $\mathcal{L}_A$-structure, such as $\mathbb{N}$, and interpreting the symbol c as some fixed number in $\mathbb{N}$. If we interpret c as $n + 1$, then the sentence φ_n^c becomes true in this "new" structure, since it holds that $n + 1 > n$.

Every formula φ_n (defined in (4.2)) is true in the standard model $\mathbb{N}$, since there is no largest natural number. The latter fact can also be expressed in the extended language as

for every n, the $\mathcal{L}_A^c$-axiom system $(\mathsf{PA} + \varphi_n^c)$ has a model.

But why would we choose this rather artificial form of expressing it? There is a crucial difference between φ_n and φ_n^c that becomes visible only when we look at all formulas simultaneously.

Let S be the set of $\mathcal{L}_A$-formulas

$$S = \{\varphi_n : n \in \mathbb{N}\}.$$

4.2. Non-standard models of Peano arithmetic

As we noted above, every sentence φ_n is true in $\mathbb{N}$ and therefore the standard model $\mathbb{N}$ is also a model of the theory PA + S. Now let us do the same with φ_n^c. Let S^c be the set of $\mathcal{L}_A^c$-formulas

$$S^c = \{\varphi_n^c : n \in \mathbb{N}\}.$$

Is $\mathbb{N}$ a model of PA+S^c? If so, there would have to be a *single* number $m \in \mathbb{N}$, the interpretation of the constant symbol c, such that $\mathbb{N} \vDash \varphi_n^c$ for all n. This number would have to be the same for all φ_n^c. Hence there would have to be a natural number greater than all natural numbers, which is not true. Note the important difference between S and S^c: To make all formulas in S true, we can choose a different witness for each φ_n, while for S^c the witness has to be *the same for every* φ_n^c, since the interpretation of a constant symbol is fixed.

Therefore, $\mathbb{N}$ cannot be turned into a model of PA + S^c by a suitable interpretation of the constant symbol c. But that does not mean that PA+S^c has no model at all. The existence of a model is in fact an easy consequence of the compactness theorem. PA + S^c has a model if and only if every finite subset of PA + S^c has a model. So let $T \subseteq$ PA + S^c be finite. T can contain two kinds of statements: axioms of PA and formulas φ_n^c. In particular, there will be *only finitely many* formulas of the second kind, say

$$c > \underline{12}, \quad c > \underline{298}, \quad c > \underline{8623}, \quad c > \underline{19191919}.$$

But for these finitely many formulas, we can easily give a model: Take $\mathbb{N}$ (which satisfies PA and, in particular, every finite subset of PA) and interpret c as 19191920, that is, one larger than the largest number occurring in any of the formulas φ_n^c of T. This gives us a model of T.

By compactness, PA + S^c has a model, say $\mathcal{N}$. $\mathcal{N}$ satisfies every axiom of PA, and it must also interpret the constant c so that every statement

$$c > \underline{n} \quad (n \in \mathbb{N})$$

is true. In other words, $\mathcal{N}$ must have an element that is larger than every natural number (or rather, every element of $\mathcal{N}$'s version of $\mathbb{N}$). Such an element is called a **non-standard number**.

The structure of non-standard models. Once the existence of non-standard models is established, one can go ahead and figure out more about the structure of non-standard models.

Induction is a crucial tool in establishing the following properties. They are not hard to prove, but may require a lot of intermediate steps. You should try to prove them, or at least sketch a proof. Details can be found the book by Kaye [39].

Properties of non-standard models of PA
- Every non-standard model has infinitely many non-standard elements (since if a is non-standard, so are $S(a), S(S(a)), \ldots$).
- The relation $<$ as defined by (4.3) defines a linear order on every model of PA with 0 as a minimal element.
- For any model $\mathcal{N}$ of PA, $\mathbb{N}^{\mathcal{N}}$ is an initial segment of $\mathcal{N}$; that is, if $c \notin \mathbb{N}^{\mathcal{N}}$ is a non-standard element of $\mathcal{N}$, then $c > n^{\mathcal{N}}$ for all $n \in \mathbb{N}$.

We can say more about the non-standard part. Suppose a is non-standard. Any non-zero element in a model of PA has an immediate predecessor, i.e. there exists b such that $S(b) = a$. Since the immediate successor of a natural number is always a natural number, it follows that b must be non-standard and has, in turn, its own non-standard predecessor. It follows that for every non-standard element a there must be non-standard elements

$$\ldots a - 3, a - 2, a - 1, a, a + 1, a + 2, a + 3 \ldots \qquad (n \in \mathbb{N}).$$

Order-wise, this means that the non-standard part contains a copy of $\mathbb{Z}$.

Furthermore, the model must also contain $a + a, a + a + a, \ldots$, i.e. $n \cdot a$ for all $n \in \mathbb{N}$. Since $a > n$ for every $n \in \mathbb{N}$ and it is easy to show by induction that in any model of PA,

$$a < b \text{ implies } a + c < b + c \text{ for all } c,$$

it follows that $a + n < 2a$ for all n. One can even show that $2a - m > a + n$ for any $m, n \in \mathbb{N}$. This means that the non-standard part

4.2. Non-standard models of Peano arithmetic

contains another copy of $\mathbb{Z}$ above the $\mathbb{Z}$-copy about a, and another $\mathbb{Z}$-copy above that, and so on. Because of the way non-standard elements interact with each other arithmetically, one can even show that between any two $\mathbb{Z}$-copies there must be another $\mathbb{Z}$-copy.

Cuts. We have seen that $\mathbb{N}$ is an initial segment of any model of PA. Formally, an **initial segment** of a model $\mathcal{N}$ is a subset I such that

$$x \in I \text{ and } y < x \Rightarrow y \in I.$$

$\mathbb{N}$ is not only an initial segment but also *closed under the successor function S*. Non-empty initial segments with this additional property are called **cuts**.

Exercise 4.15. Let $\mathcal{N}$ be a model of PA and let $a \in \mathcal{N}$. Show that

$$a^{\mathbb{N}} = \{x : x < a^n \text{ for some } n \in \mathbb{N}\}$$

is a cut in $\mathcal{N}$.

A very interesting (and most important) fact about cuts is that they are very hard to describe. This assertion seems to contradict the example we gave above: $\mathbb{N}$ is always a cut, and it is very easy to describe!

While, of course, $\mathbb{N}$ is easy to describe in a certain sense, this depends on our capability to look at a model of PA "from outside". Imagine you are being tossed into a non-standard model of PA, and land on some element a. You know nothing about the model other than that it is a model of PA. Can you tell, by looking at the elements around you, whether you are in the standard or non-standard part? In other words, is there a property that distinguishes the standard elements from the non-standard elements?

To make this question more precise, we have to say what we mean by "property".

Definition 4.16. Let $\mathcal{L}$ be a language, and let $\mathcal{M}$ be an $\mathcal{L}$-structure. A set $A \subseteq M^n$ is **definable** if there exists an $\mathcal{L}$-formula $\varphi(x_1, \ldots, x_n)$ with n free variables $x_1, \ldots, x_n$ such that

$$(a_1, \ldots, a_n) \in A \quad \Leftrightarrow \quad \mathcal{M} \vDash \varphi[a_1, \ldots, a_n].$$

We view $\varphi(x_1,\ldots,x_n)$ as a *property* and view the set A defined by φ as the set of elements having the property φ.

Example 4.17.

(1) In any group G, the *centralizer* can be defined via
$$\varphi(x) \equiv \forall y\, (y \circ x = x \circ y).$$

(2) In any $\mathcal{L}_A$-structure, the set of even elements is defined by
$$\varphi(x) \equiv \exists y\, (x = y + y).$$

In the second example, if we consider the standard model of PA, we get exactly the *even natural numbers*. But in a non-standard model, the definition will also include non-standard even numbers—any number that satisfies the definition.

Can we define only the even *standard* numbers? This leads us back to the original question: why a cut cannot be easily described.

While the set of even standard numbers is not itself a cut, it can be easily turned into one by closing it downwards under $<$. If we can define the even standard numbers, say by a formula $\varphi_E(x)$, we can define the cut $\mathbb{N}$ given by all standard numbers through
$$x \in \mathbb{N} \quad \Leftrightarrow \quad \exists y\, (\varphi_E(y) \wedge x < y).$$

This means that if the even standard numbers are definable, so is the set of all standard numbers.

Proposition 4.18. *Suppose I is a cut in a non-standard model $\mathcal{N}$ of PA, and suppose I is proper, that is, I is not all of $\mathcal{N}$. Then I is not definable.*

Proof. Assume for a contradiction that I is definable via a formula φ, that is,
$$a \in I \quad \Leftrightarrow \quad \mathcal{N} \vDash \varphi[a].$$
Since $I \neq \varnothing$ and is closed downwards under $<$, $0 \in I$ and thus $\varphi[0]$ holds in $\mathcal{N}$. Furthermore, if $a \in I$, then $S(a) \in I$, because I is a cut. This means in turn that
$$\forall x\, (\varphi(x) \rightarrow \varphi(S(x)))$$

4.2. Non-standard models of Peano arithmetic

holds in $\mathcal{N}$. But $\mathcal{N}$ satisfies the induction axiom corresponding to φ, and we just established that the antecedent of the axiom

$$\varphi(0) \land \forall x \, (\varphi(x) \to \varphi(S(x)))$$

is true in $\mathcal{N}$. This means that, as a consequence of the induction axiom,

$$\mathcal{N} \vDash \forall x \, \varphi(x).$$

In other words, $I = \mathcal{N}$, contradicting our assumption that I is proper. □

This proposition tells us that there can be no (definable) property that distinguishes between the elements of a cut and the elements of its complement. In particular, there is no definable property that expresses "*x is a standard number*".

You may raise the following objection:

Numbers in the standard part have at most finitely many predecessors, while numbers in the non-standard part have infinitely many. Is this not a definable property?

The problem here lies with "finitely/infinitely many". How would you write down a formula for this? The above proof in fact shows that we cannot express this in the language of arithmetic.

Since $\mathbb{N}$ is a cut in any model of PA, it follows that any definable property that holds for infinitely many standard numbers must hold for a non-standard number, too. This is called **overspill**.

Corollary 4.19 (Overspill). *Suppose $\mathcal{M}$ is a non-standard model of* PA *and $\varphi(x)$ is an $\mathcal{L}_A$-formula. If there exist infinitely many $n \in \mathbb{N}$ such that $\mathcal{M} \vDash \varphi[n]$, then there exists a $c \in \mathcal{M} \setminus \mathbb{N}$ such that $\mathcal{M} \vDash \varphi[c]$.*

Proof. If no such c existed, we could define $\mathbb{N}$ in $\mathcal{M}$ since

$$\mathbb{N} = \{a \in M : \text{exists } b > a \; \mathcal{M} \vDash \varphi[b]\}.$$

□

4.3. Ramsey theory in Peano arithmetic

All proofs in this book so far have *not* been fully formalized proofs in the sense of Definition 4.2. Instead, we gave what one might call "semi-formal" proofs. We used natural language mixed with mathematical notation and calculations to describe the essential deductive steps.

What would fully formal proofs look like? Let us consider the finite Ramsey theorem, Theorem 1.31. Is this formally provable in Peano arithmetic? If we look at the statement of Theorem 1.31, we notice that it does not immediately translate to a formal sentence in the language of arithmetic $\mathcal{L}_\mathcal{A}$, since it uses symbols not available in $\mathcal{L}_\mathcal{A}$. In a first step, we would have to show how to formulate this statement using only $S, +, \cdot, 0$. This already proves rather tedious. Let us first restate the theorem, this time without using the arrow notation.

Theorem 4.20. *For any natural numbers k, p, and r there exists a natural number N such that for every function $c : [N]^p \to \{1, \ldots, r\}$ there exists a subset H of $\{1, \ldots, N\}$ of cardinality k and a number j, $1 \le j \le r$, such that for all p-element subsets $\{a_1, \ldots, a_p\}$ of H, $c(\{a_1, \ldots, a_p\}) = j$.*

This is already a complicated statement, but it is far from being a statement in the language of arithmetic. For example, we have no symbols to speak about *subsets* or *functions* (other than $S, +, \cdot$). We would have to show that these notions can be defined over PA. In (4.3) we defined the < symbol by giving a formula that could be "plugged in" every time the symbol occurs in a formula. Can we do something similar for subsets? The problem is that sets of numbers are of a different *type* than numbers.

The solution to this problem introduces a fundamental concept in mathematical logic, *Gödelization*. In a nutshell, this is the process of *coding objects by natural numbers*, so that we can use $\mathcal{L}_\mathcal{A}$-formulas to express facts about them.

Coding functions. We have already encountered a coding function for *pairs*, Cantor's pairing function $\langle x, y \rangle$, in Section 2.5. To code

4.3. Ramsey theory in Peano arithmetic

arbitrary sequences, we could iterate the pairing function by letting $\rho_2(x,y) = \langle x,y \rangle$ and define $\rho_{n+1} : \mathbb{N}^{n+1} \to \mathbb{N}$ by

$$\rho_{n+1}(x_1, \ldots, x_{n+1}) = \langle x_1, \rho_n(x_1, \ldots, x_n) \rangle.$$

This, however, would give us a whole family (π_n) of functions, one for every length.

It is possible to code sequences of natural numbers by a single function, independent of their lengths. Let

(4.4) $$\rho\big((a_1, \ldots, a_k)\big) = 2^{a_1+1} \cdot 3^{a_2+1} \cdots p_k^{a_k+1},$$

where p_k is the kth prime number. The uniqueness of prime decomposition ensures that this is a one-to-one function, so every code number represents a unique sequence. Not every number will be a code number in this way: for example, $45 = 3^2 \cdot 5$ is not, since every code number has 2 as a factor (note that we add 1 to every exponent in the definition above). But given a number, we can effectively (meaning by using a computer program) recognize whether a number is the code of a sequence and, if so, compute the coded sequence. We could, therefore, effectively renumber the codes so that every number represents a code: Find the least number that represents a code and assign it code 0, then find the second smallest number that represents a code and assign it 1, and so forth.

This coding function works very well, but it has one big caveat: In the language of arithmetic $\mathcal{L}_A$, we do not have a symbol for exponentiation, so we cannot directly express the coding function above in $\mathcal{L}_A$.

Gödel found a coding scheme that works in formal arithmetic (in particular, it works in PA).

For this, we need the *remainder* function:

$\text{rem}(x, y)$ = the remainder when x is divided by y (as integers).

Of course, $\text{rem}(x, y)$ is not a symbol of $\mathcal{L}_A$ either, but we can define it (or rather, its graph) easily:

(4.5) $$\text{rem}(x,y) = z \iff 0 \leq z < y \wedge \exists c \leq x \, (x = cy + z).$$

Similar to <, we can now use rem in $\mathcal{L}_A$-formulas. We write
$$x \equiv a \pmod{y} \text{ if } \text{rem}(x-a, y) = 0, \text{ i.e. } y \text{ divides } x - a,$$
and say that x *is congruent to a modulo y*. If $a < y$, then $x \equiv a \pmod{y}$ implies $\text{rem}(x, y) = a$. We further say that two integers x and y are *relatively prime* if they have no common prime factors, that is, if $\gcd(x, y) = 1$.

Theorem 4.21 (Chinese remainder theorem). *Let $m_0, \ldots, m_{n-1}$ be pairwise relatively prime integers. Given any sequence of integers $a_0, \ldots, a_{n-1}$, there exists an integer x such that for $i = 0, \ldots, n-1$,*
$$x \equiv a_i \pmod{m_i}.$$

In other words, for pairwise relatively prime integers, a system of equations of multiple congruences always has a solution. There are many proofs known for the Chinese remainder theorem. Rautenberg [**54**] and Smoryński [**62**] give elementary proofs that can easily be formalized in PA.

Gödel's idea was to use a solution x as a code for the sequence $(a_0, \ldots, a_{n-1})$. If we can find a suitable sequence of pairwise relatively prime $m_0, \ldots, m_{n-1}$ with $m_i > a_i$ for $0 \leq i < n$, then to decode x we would only have to determine the remainder of x divided by m_i for each i, a very simple arithmetic operation.

Lemma 4.22. *For any integer $l > 0$, the integers*
$$1 + l!, \ 1 + 2 \cdot l!, \ldots, 1 + l \cdot l!$$
are pairwise relatively prime.

Proof. Suppose there exists a prime p that divides both $1 + i \cdot l!$ and $1 + j \cdot l!$ for $i < j$. Then p also divides
$$(1 + j \cdot l!) - (1 + i \cdot l!) = (j - i) \cdot l!.$$
Since p is a factor of $1 + i \cdot l!$, it cannot be a factor of $l!$. As $l!$ has all numbers $\leq l$ as its factors, it must hold that $p > l$. On the other hand, since p is prime, it follows that p divides $j - i$. (Recall that p is prime if and only if whenever p divides ab, it must divide at least one of a and b.) This would imply that
$$1 < p \leq j - i < l,$$
which contradicts $p > l$. □

4.3. Ramsey theory in Peano arithmetic

The lemma tells us that we can generate sequences of pairwise relatively prime numbers of arbitrary length l. If we combine this with the Chinese remainder theorem, we get the desired coding function.

Definition 4.23 (Gödel β-function). For arbitrary natural numbers c, d, and i, let
$$\beta(c, d, i) = \text{rem}(c, 1 + (i+1)d).$$

Theorem 4.24. *Let $a_0, \ldots, a_{n-1}$ be a finite sequence of natural numbers. There exist c and d such that for each $i = 0, \ldots, n-1$,*
$$\beta(c, d, i) = a_i.$$

Proof. The main point is to choose a large enough l and then apply Lemma 4.22. Let $a = \max\{a_0, \ldots, a_{n-1}\}$ and put $l = an$. By Lemma 4.22, the numbers
$$1 + l!, \ldots, 1 + l \cdot l!$$
are pairwise relatively prime. By the Chinese remainder theorem, there exists c such that for $i = 0, \ldots, n-1$,
$$c \equiv a_i \pmod{1 + (i+1)l!}.$$
Since
$$a_i \leq a \leq an = l \leq l! < 1 + (i+1)l!,$$
we have that $\text{rem}(c, 1 + (i+1)l!) = a_i$, and thus, if we let $d = l!$, $\beta(c, d, i) = a_i$. □

We can view the numbers c and d together as a code for the sequence $a_0, \ldots, a_{n-1}$. Using the Cantor pairing function, we can combine them into a single number $\langle c, d \rangle$.

We demonstrate the usefulness of the β-function by defining exponentiation. In Definition 4.16, we laid out what it means for a subset of a structure to be definable. As any function can be identified with its graph, we can extend this definition to functions as follows.

Definition 4.25. Suppose $\mathcal{M}$ is an $\mathcal{L}$-structure and f is a function from M^k to M. f is **definable** (in $\mathcal{M}$) if there exists an $\mathcal{L}$-formula $\varphi(x_1, \ldots, x_k, y)$ such that for all $a_1, \ldots, a_k, b \in M$,
$$f(a_1, \ldots, a_k) = b \iff \mathcal{M} \models \varphi[a_1, \ldots, a_k, b].$$

156   4. Metamathematics

Proposition 4.26. *The function* $(x,y) \mapsto x^y$ *is definable in* $\mathbb{N}$ *(in the language* $\mathcal{L}_A$*).*

Proof. For any $x, y, z \in \mathbb{N}$, $x^y = z$ if and only if the following formula holds in $\mathbb{N}$:

(4.6) $\quad \exists c \exists d \, \big[\beta(c,d,0) = S(0) \; \wedge$
$$\forall v < y \; \beta(c,d,S(v)) = \beta(c,d,v) \cdot x \; \wedge \; \beta(c,d,y) = z \big].$$

$\square$

The formula in (4.6) expresses that the β-code $\langle c, d \rangle$ codes a sequence that represents a correct computation of x^y, by iterating multiplication by x, starting from $x^0 = 1$ and ending at $x^y = z$. In Chapter 3, we defined new functions φ_{n+1} by iterating φ_n. Using the β-function, we can code these definitions as well and define the functions φ_n via formulas. More generally, with the help of the β-function it is not hard to show that

(1) the basic primitive recursive functions Zero, S, and P_n^i are definable;

(2) if f and g are definable, then the composition $f \circ g$ and the function h obtained by recursion from f and g are definable.

These basic functions and elementary operations were defined in Section 3.3, and we can now see the following.

Proposition 4.27. *Every primitive recursive function* $f : \mathbb{N}^k \to \mathbb{N}$ *is definable in* $\mathbb{N}$ *in the language of arithmetic* $\mathcal{L}_A$.

For a full proof, see for example [**13**] or [**54**]. The proposition gives us a first glimpse of a powerful connection between logic and computation, a connection that will feature prominently in the next section.

Coding number-theoretic statements in PA. Coding via the β-function has some peculiarities: Every sequence has multiple β-codes; and the β-function is defined for *every* input (c, d, i), not just the "intended" i, which means that the *length* of the sequence coded by $\langle c, d \rangle$ is not clearly defined.

4.3. Ramsey theory in Peano arithmetic

Let us formally define the decoding function $\text{decode}(x, i)$ as follows:

Find the unique c and d such that $x = \langle c, d \rangle$ and let
$$\text{decode}(x, i) = \text{rem}(c, 1 + (i + 1)d).$$

Many authors use $(x)_i$ to denote $\text{decode}(x, i)$.

To resolve the length issue, we will assume that the 0-entry of any sequence is its length. We define
$$\text{length}(x) = (x)_0 = \text{decode}(x, 0).$$

Exercise 4.28. Show that both decode and length are definable in $\mathbb{N}$.

Using codes, we can express relations between finite sets of numbers as arithmetic relations between codes. For example, we can express "x codes a finite set of cardinality k" as

$$\text{length}(x) = k \ \wedge \ \forall i, j \left((1 \leq i, j \leq k \ \wedge \ i < j) \right.$$
$$\left. \Rightarrow \text{decode}(x, i) < \text{decode}(x, j) \right).$$

As you see, we require that the elements of the set be coded in increasing order.

We will use the predicate $\text{set}(x, k)$ to denote the formula above. As functions on $\mathbb{N}$ correspond to sequences of values, we can use codes to represent functions defined on finite initial segments of $\mathbb{N}$, too. Let the predicate $\text{func}(x, k)$ be defined as

$$\text{length}(x) = 2 \ \wedge \ \text{length}(\text{decode}(x, 1)) = k \ \wedge$$
$$\text{length}(\text{decode}(x, 2)) = k \ \wedge \ \text{set}(\text{decode}(x, 1)),$$

which expresses the property that x codes two sequences, both of length k, and the first sequence is a set (the domain of the function). If $\text{func}(x, k)$, we define, for $i \leq k$,

$$\text{arg}(x, i) = \text{decode}(\text{decode}(x, 1), i),$$
$$\text{val}(x, i) = \text{decode}(\text{decode}(x, 2), i),$$

which returns the ith argument and the value of the function at the ith argument, respectively.

We can finally state the finite Ramsey theorem fully formalized (if we expand the abbreviations above) in the language of arithmetic:

$$\forall k, p, r \exists N \forall f \bigg[$$
$$\exists l \bigg(\mathrm{func}(f, l) \wedge$$
$$\forall i \leq l \, \big(\mathrm{set}(\arg(f, i), p) \wedge$$
$$\forall j \leq p (\mathrm{decode}(\arg(f, i), j) \leq N) \big) \wedge$$
$$\forall y \, \big((\mathrm{set}(y, p) \wedge \forall j \leq p (\mathrm{decode}(y, j) \leq N))$$
$$\Rightarrow \exists j (\arg(f, j) = y) \big)$$
$$\forall j \leq l \, (\mathrm{val}(f, j) \leq r) \bigg)$$
$$\Longrightarrow$$
$$\exists z, j \bigg(j \leq r \wedge \mathrm{set}(z, k) \wedge$$
$$\forall i \leq k \, (\mathrm{decode}(z, i) \leq N) \wedge$$
$$\forall i \leq k \, \forall m \leq l \big(\arg(f, m) = \mathrm{decode}(z, i) \Rightarrow \mathrm{val}(f, m) = j \big) \bigg) \bigg]$$

The upper half of the formula expresses that f is a function from $[N]^p$ to $\{1, \ldots, r\}$, while the lower part formalizes the existence of a homogeneous subset of size k.

As we see, formalization in PA is a rather tedious business. And we have only formalized the statement of Ramsey's theorem itself so far. What we would really like to affirm is that the *proof* can be formalized, too.

Recall from Definition 4.2 that a formal proof is a sequence of formulas such that each formula is either an axiom or obtained from previous formulas through an application of the deduction rule. Is this possible for the proof of Ramsey's theorem, Theorem 1.31? It is. We hope the reader will take our word for it, as we will not present such a formalization here. Of course, we encourage everyone to carry out at least a few steps of such a formal proof herself. A good "compromise"

4.4. Incompleteness

would be to go over the essential steps in the proof and check that they can be established simply by using the basic properties of addition and multiplication via induction. (If you are willing to study a little more mathematical logic, you will acquire some nice tools such as *upward absoluteness* that make this a lot easier.)

We can also point to the rich efforts of a community of mathematicians aimed at formalizing proofs of major results so that the correctness of the proofs can subsequently be checked by a computer. This has been done multiple times for Ramsey's theorem; see for example [55].

Which other number-theoretic theorems can be proven in Peano arithmetic? It turns out that in the vast majority of cases, if you can formalize the statement in arithmetic, then it is provable in PA. Euclid's theorem on the infinitude of primes, van der Waerden's theorem (Theorem 3.3), the law of quadratic reciprocity—all are provable in PA.[4] This is good evidence that PA is quite strong as a formal system and captures most of elementary number theory.

On the other hand, one might ask:

Are there true statements about the natural numbers that cannot *be proved in* PA*?*

This question spurred some of the greatest achievements in mathematical logic in the 20th century. Along the way, the optimism that mathematics could be put on a completely solid foundation was shattered. But it also produced some beautiful mathematics, in which Ramsey theory played no small part.

4.4. Incompleteness

The natural numbers $\mathbb{N}$ (just like any other $\mathcal{L}_A$-structure) have the property that any $\mathcal{L}_A$-sentence is either true or false in $\mathbb{N}$, and in the latter case this means that the negation of the sentence must be true.

$$\text{For any sentence } \sigma, \text{ either } \mathbb{N} \vDash \sigma \text{ or } \mathbb{N} \vDash \neg\sigma.$$

If a theory has this property, it is called *complete*.

---

[4] A notable uncertainty at the time this book was written concerns Wiles's proof of Fermat's last theorem, but there is optimism among experts that it can be formalized in PA as well.

4. Metamathematics

Definition 4.29. A theory T is **complete** if for every sentence σ,

$$\text{either } T \vdash \sigma \text{ or } T \vdash \neg\sigma.$$

Note that according to this definition, complete theories are automatically consistent.[5] It is important to distinguish between completeness of *theories* and the completeness of a *logic* (via a proof system, as discussed in Section 4.1).

Exercise 4.30. Show that the set of consequences of a complete theory T,

$$T_\vDash = \{\sigma \colon T \vDash \sigma\},$$

is *maximally consistent*, that is, $T_\vDash$ is consistent and no proper extension of $T_\vDash$ is consistent.

Exercise 4.31. Given a language $\mathcal{L}$ and an $\mathcal{L}$-structure $\mathcal{M}$, the $\mathcal{L}$-*theory of* $\mathcal{M}$ is defined as

$$\text{Th}(\mathcal{M}) = \{\sigma \colon \sigma \text{ is an } \mathcal{L}\text{-sentence and } \mathcal{M} \vDash \sigma\}.$$

Show that $\text{Th}(\mathcal{M})$ is always a complete theory.

Completeness itself is not a mathematical virtue. Many axiom systems, such as the group axioms G, describe very general structures. Groups occur in so many different forms and settings that it would be very surprising if all groups would satisfy exactly the same statements in the language of group theory. Consider, for example, the statement

$$\forall x, y \, (x \circ y = y \circ x).$$

The statement says that a group is Abelian, i.e. the group operation commutes. G has no opinion about this statement, since it neither proves the statement nor disproves it (i.e. proves its negation). Some groups are Abelian, others not. Hence the theory of groups is incomplete.

Exercise 4.32. Give some other examples of incomplete theories.

---

[5] Some authors define completeness using a non-exclusive *or*. Completeness in the sense of Definition 4.29 would then be equivalent to being complete *and* consistent.

4.4. Incompleteness

The situation is different if we want to axiomatically describe a *single* mathematical structure. PA is an attempt to capture number-theoretic properties about the natural numbers expressible in first-order logic.

Since $\mathbb{N}$ is a model of PA, everything PA proves must be true in $\mathbb{N}$. And if PA were complete, everything that is true in $\mathbb{N}$ would be provable in PA: If $\mathbb{N} \vDash \sigma$ but PA $\nvdash \sigma$, then, since PA is complete, PA $\vdash \neg\sigma$ and hence $\mathbb{N} \vDash \neg\sigma$, a contradiction. In other words, if PA were complete, PA would prove *exactly* the true statements about $\mathbb{N}$. In the notation of Exercises 4.30 and 4.31, we would have PA$_\vDash$ = Th($\mathbb{N}$).

This would be strong evidence of PA being the "right" first-order axiomatization of arithmetic. While we cannot avoid the existence of non-standard models (even if PA is complete), it would tell us that we can axiomatize the theory of $\mathbb{N}$ by means of a simple (though infinite) axiom system.

So we arrive at the following question:

Is PA *complete?*

It turns out that if PA were complete, we would be able to solve problems via computer that are demonstrably not solvable. In particular, we would be able to solve the *halting problem*. Therefore, PA is not complete. We will try to elaborate on this, returning, as promised, to the fascinating connection between logic and computability.

The Entscheidungsproblem. In 1928, David Hilbert [32] asked whether there is an algorithm that, given as input a formula of first-order logic, decides whether this formula is *universally valid*, that is, holds in every structure.[6] This question became known as the *Entscheidungsproblem* (decision problem).

We can ask a similar question with respect to truth in $\mathbb{N}$:

Is there an algorithm that, given as input an $\mathcal{L}_\mathcal{A}$-sentence σ, decides whether it holds in $\mathbb{N}$ or not (that is, whether $\sigma \in \text{Th}(\mathbb{N})$ or $\neg\sigma \in \text{Th}(\mathbb{N})$)?

---

[6] An example of such a formula would be $\forall x\, (x = x)$.

The proof that a general solution to the Entscheidungsproblem, for general validities and for the theory of the natural numbers as formulated above, is impossible marks one of the great milestones of mathematical logic as well as the beginning of modern computer science.

How can one show that there is no computer program that can perform the task above, or, in general, any given task? Of course, one would have to agree on a mathematically rigorous definition of what constitutes being *solvable by a computer*. In this day and age, where computers, algorithms, and programming languages are ubiquitous and part of everyday life, everyone has some idea of what this means, but when the Entscheidungsproblem was first formulated in the 1920s, things were different.

Computable functions and Gödelization. Intuitively, a function $f : \mathbb{N}^n \to \mathbb{N}$ is **computable** if there exists an algorithm that on input $\bar{x}$ follows an effective, deterministic, finite procedure and eventually outputs $f(\bar{x})$. In his groundbreaking work in 1936 [66], Alan Turing introduced a model of an abstract machine with which he tried to capture both the notion of an "algorithm" and the "calculation procedure" mathematically. This machine model now carries his name: the **Turing machine**. As a Turing machine is a rigorously specified mathematical concept, one can formally define a function $f : \mathbb{N} \to \mathbb{N}$ to be computable if there exists a Turing machine M such that for any x, M's computation on input x terminates after a finite number of steps and outputs $f(x)$. A set $A \subseteq \mathbb{N}$ is computable if its characteristic function is:

$$\chi_A(n) = \begin{cases} 1 & \text{if } n \in A, \\ 0 & \text{if } n \notin A. \end{cases}$$

We will not give a definition of Turing machines here, but you can find one in any good textbook on computability theory (such as [63]). However, we point out two central features of Turing machines which they share with modern computation devices (programming languages such as Python, C++, or Java, and hardware on which programs can be executed):

4.4. Incompleteness

- A Turing machine program is a finite list of instructions, which are finite strings over a finite alphabet of symbols.

- During the execution of a Turing machine program, at any time the current configuration of the machine can be described in a finite way. Think of a *snapshot* of a computer running a program—pretty much what happens when you debug code. It would consist of the current contents of memory and hard drive, the program line that is currently being executed, the current values of variables, and so on; it could all be "dumped" into one big file, say "snapshot.txt".

In fact, one could define computable functions using Python, C++, Java, or whatever your favorite programming language is. Turing machines have the advantage that they are, on the one hand, based on a very simple model that is amenable to rigorous mathematical analysis. On the other hand, they are just as powerful as any modern computer and any modern programming language, in the sense that they give rise to the same family of computable functions.

The notion of Turing computability also turns out to be equivalent to a number of other formalizations of computability that have been suggested over the years, from *register machines* to the λ-*calculus*. They all define the same notion of what it means for a problem to be algorithmically solvable. This further confirmed that Turing's analysis of the intuitive notion of algorithm was adequate, and gave rise to the **Church-Turing thesis**. The thesis states that any problem that is informally solvable by an algorithm is solvable by a Turing machine (or any of the notions equivalent to it). We will frequently appeal to this thesis in what follows by describing informal algorithms when proving that a certain function is computable.

Every primitive recursive function (see Definition 3.16) is computable: The three basic functions are computable. Both composition and recursion can be simulated by a Turing machine, and hence any function that is the result of a finite number of applications of basic functions, composition, and recursion is computable. There are, as we saw in Corollary 3.26, computable functions that are not primitive recursive, such as the Ackermann function $\varphi(x, y, n)$. It is in fact not very hard to write a Python, C, or Java program to compute φ (though it would take a very long time to run it).

A program (Turing machine or other) is just a finite string of symbols. Whenever we have a finite string of symbols over a finite (or countably infinite) alphabet, we can devise an algorithmic coding scheme that assigns each program a natural number, its **Gödel number**. The basic idea is very similar to the Gödel numbering of finite sequences of natural numbers that we introduced in the previous section. In the following, we develop a Gödel numbering for formulas of first-order arithmetic. Formulas are, after all, finite sequences over a countable alphabet, too.

The basic idea is to assign numbers to the basic symbols and then, as before, use products and powers of prime numbers to encode sequences of symbols. The usual notation for Gödel numbers of formulas is $\ulcorner \varphi \urcorner$. First we define

$$\ulcorner 0 \urcorner = 1,$$

$$\ulcorner x_i \urcorner = 2^{i+1},$$

which assigns a Gödel number to all variables and the constant symbol 0. Next, we assign Gödel numbers to all terms, i.e. expressions that can be obtained by applying the operations $S, +, \cdot$ to variables, constants, and other terms. Suppose s and t have already been assigned Gödel numbers. Then we put

$$\ulcorner S(t) \urcorner = 3 \cdot 7^{\ulcorner t \urcorner},$$

$$\ulcorner s+t \urcorner = 3^2 \cdot 7^{\ulcorner s \urcorner} \cdot 11^{\ulcorner t \urcorner},$$

$$\ulcorner s \cdot t \urcorner = 3^3 \cdot 7^{\ulcorner s \urcorner} \cdot 11^{\ulcorner t \urcorner}.$$

Finally, we assign Gödel numbers to formulas:

$$\ulcorner s = t \urcorner = 5 \cdot 7^{\ulcorner s \urcorner} \cdot 11^{\ulcorner t \urcorner},$$

$$\ulcorner \neg \varphi \urcorner = 5^2 \cdot 7^{\ulcorner \varphi \urcorner},$$

$$\ulcorner \varphi \wedge \psi \urcorner = 5^3 \cdot 7^{\ulcorner \varphi \urcorner} \cdot 11^{\ulcorner \psi \urcorner},$$

$$\ulcorner \exists x_i(\varphi) \urcorner = 5^4 \cdot 7^{i+1} \cdot 11^{\ulcorner \varphi \urcorner}.$$

4.4. Incompleteness

For example, the formula $\exists x_0 (x_0 + x_1 = x_0 \cdot x_2)$ has the Gödel number

$$5^4 \cdot 7^1 \cdot 11^{\ulcorner x_0 + x_1 = x_0 \cdot x_2 \urcorner} = 5^4 \cdot 7 \cdot 11^{5 \cdot 7^{\ulcorner x_0 + x_1 \urcorner} \cdot 11^{\ulcorner x_0 \cdot x_2 \urcorner}}$$

$$= 5^4 \cdot 7 \cdot 11^{5 \cdot 7^{3^2 \cdot 7^2 \cdot 11^{2^2}} \cdot 11^{3^3 \cdot 7^2 \cdot 11^{2^3}}}.$$

Exercise 4.33. Compute the Gödel numbers of $0 + x_5 = x_5$ and $\neg \exists x_1 (S(x_1) = 0)$.

There are, of course, other ways to devise a Gödel numbering of $\mathcal{L}_A$-formulas. The important properties of any Gödel numbering for formulas are the following:

- The mapping $\varphi \mapsto \ulcorner \varphi \urcorner$ from formulas to Gödel numbers is injective.
- Given a number $m \in \mathbb{N}$, we can algorithmically check whether m is the Gödel number of some formula φ.
- If m is the Gödel number of an $\mathcal{L}_A$-formula φ, we can compute φ from m.

In the encoding defined above, all properties are ensured through the effectiveness and uniqueness of the prime decomposition of a number.

Exercise 4.34. Devise (informally) a Gödel numbering of programs in your favorite programming language.

As before, we can also rearrange the Gödel numbers so that every number becomes the Gödel number of a formula, that is, the mapping $\varphi \mapsto \ulcorner \varphi \urcorner$ is one-to-one and onto. Similarly, we can do this for a Gödel numbering of Python programs or Turing machines.

So let us assume now that we have fixed a Gödel numbering

$$M_0, M_1, M_2, \ldots$$

of Turing machines. Here, M_i denotes the Turing machine with Gödel number i.

As you may have learned the hard way while learning how to program, a computer program may fail to terminate. For instance, you could implement a WHILE-loop whose exit condition is never met. The same holds true for Turing machines. They may fail to terminate on certain inputs, while for other inputs they finish and output a result after finitely many steps.

Using the Gödel numbering of Turing machines, we can define the following set of natural numbers:

$$K = \{x : M_x \text{ halts after finitely many steps on input } x\}.$$

The set K is known as the **halting problem** for Turing machines. There is in fact nothing special about Turing machines here. Given a suitable Gödel numbering scheme, one can define it for other machine models or programming languages, too.

The following is one of the fundamental results of the theory of computability.

Theorem 4.35 (Unsolvability of the halting problem; Turing [66]). *The halting problem for Turing machines is unsolvable, that is, the set K defined above is not computable.*

Proof. We identify M_i with the partial function it computes and write $M_i(x) \downarrow = y$ if on input x, M_i halts after finitely many steps and outputs y; we write $M_i(x) \uparrow$ if on input x, M_i does not halt.

Assume for a contradiction that K is computable. Then there exists a Turing machine M that computes the characteristic function of K. Let d be an index for M, i.e. d is such that $M = M_d$. Then

$$M_d(x) = \begin{cases} 1, & \text{if } M_x(x) \downarrow, \\ 0, & \text{if } M_x(x) \uparrow. \end{cases}$$

Using M_d we can define its "evil twin" $\widetilde{M}$: On input x we first run M_d on input x. If $M_d(x) = 1$, we send $\widetilde{M}$ into an infinite loop (that is, it will not halt on input x). If $M_d(x) = 0$, we terminate the program and output 1. Hence we have

$$\widetilde{M}(x) = \begin{cases} \uparrow & \text{if } M_d(x) = 1, \\ 1 & \text{if } M_d(x) = 0. \end{cases}$$

Note that $\widetilde{M}$ "swaps" the halting behavior of M_x: If $M_x(x) \downarrow$, then $M_d(x) = 1$ and hence $\widetilde{M}(x) = \uparrow$. If $M_x(x) \uparrow$, then $M_d(x) = 0$ and thus $\widetilde{M}(x) = \downarrow 1$.

4.4. Incompleteness

Since $\widetilde{M}$ is a machine, it has an index, say $\widetilde{M} = M_e$. Now comes the crucial question: *Does $\widetilde{M}$ halt on input e?*

We have

$$\widetilde{M}(e) \downarrow \;\Rightarrow\; M_e(e) \downarrow \;\Rightarrow\; M_d(e) = 1 \;\Rightarrow\; \widetilde{M}(e) \uparrow$$

and

$$\widetilde{M}(e) \uparrow \;\Rightarrow\; M_e(e) \uparrow \;\Rightarrow\; M_d(e) = 0 \;\Rightarrow\; \widetilde{M}(e) \downarrow,$$

which is a contradiction. It follows that M_d cannot exist. □

One can paraphrase this theorem as follows: It is impossible to write the "ultimate, ultimate debugger", a program that inspects any other program and determines correctly whether, on a given input, this program terminates after finitely many steps or runs forever.

Seemingly unrelated at first sight, the unsolvability of the halting problem also implies the undecidability of the theory of first-order arithmetic.

Theorem 4.36 (Unsolvability of the Entscheidungsproblem for $\mathbb{N}$; Turing [**66**], Church [**6**,**7**]). *The function g defined as*

$$g(\ulcorner \sigma \urcorner) = \begin{cases} 1 & \text{if } \mathbb{N} \vDash \sigma, \\ 0 & \text{if } \mathbb{N} \nvDash \sigma \end{cases}$$

is not computable.

The reason for this is that we can express facts about Turing machine computations as formulas of first-order arithmetic. In the previous section, we stated Gödel's result that every primitive recursive function is definable over $\mathbb{N}$ (Proposition 4.27). Kleene extended this to all computable functions and relations.

Theorem 4.37 (Kleene; see [**63**]). *There exists an $\mathcal{L}_A$-formula*

$$\Psi(x_0, x_1, x_2)$$

such that

$$\mathbb{N} \vDash \Psi[e, a, b] \;\Leftrightarrow\;$$

the eth Turing machine halts on input a and outputs b.

If the function g defined above were computable, K would be, too, since
$$x \in K \iff g(\ulcorner \exists y\, \Psi(\underline{x},\underline{x},y) \urcorner) = 1;$$
hence g cannot be computable. (Recall that $\underline{x}$ is the constant term representing the number x defined in Section 4.2.) Here we also see why it is important that a Gödel numbering is effective.

Finally, the undecidability of the theory of $\mathbb{N}$ implies that PA is incomplete (a weak form of Gödel's first incompleteness theorem).

Theorem 4.38. *Peano arithmetic is incomplete, that is, there exists an $\mathcal{L}_{\mathcal{A}}$-sentence σ such that*
$$neither \ \mathsf{PA} \vdash \sigma \ nor \ \mathsf{PA} \vdash \neg\sigma.$$

Proof. Assume for a contradiction that PA is complete, i.e., for every $\mathcal{L}_{\mathcal{A}}$-sentence σ,
$$either \ \mathsf{PA} \vdash \sigma \ or \ \mathsf{PA} \vdash \neg\sigma.$$
We claim that in this case the function
$$P(x) = \begin{cases} 0 & \text{if } x \text{ is the Gödel number of a sentence } \sigma \text{ and } \mathsf{PA} \vdash \sigma, \\ 1 & \text{if } x \text{ is the Gödel number of a sentence } \sigma \text{ and } \mathsf{PA} \vdash \neg\sigma, \\ 2 & \text{if } x \text{ is not the Gödel number of any sentence} \end{cases}$$
is computable. We can algorithmically check whether a number is the Gödel number of an $\mathcal{L}_{\mathcal{A}}$-sentence, so in case it is not, we can terminate the algorithm right away and output 2. In the case where x is the Gödel number of a sentence σ, we start the following search procedure:

[1] Put $a = 1$.
[2] Interpret a as a code of a finite sequence of natural numbers $(a_1, \ldots, a_n)$ as defined in (4.4).
[3] Check whether each a_i, $i = 1, \ldots, n$, is the Gödel number of an $\mathcal{L}_{\mathcal{A}}$-formula. If not, put $a := a + 1$ and go to [2].
[4] Check if the sequence of formulas coded by $(a_1, \ldots, a_n)$ represents a valid proof of σ or $\neg\sigma$ from PA. If this is not the case, put $a := a + 1$ and go to [2].
[5] If $(a_1, \ldots, a_n)$ codes a proof of σ, terminate and return 0; if it codes a proof of $\neg\sigma$, terminate and return 1.

4.4. Incompleteness

There are two important points. First, checking whether a sequence of formulas is a PA-proof of σ (or $\neg\sigma$) can be done algorithmically. By the definition of a formal proof (Definition 4.2), we are able to check whether a formula is an axiom (of PA or is a logical axiom) or the result of a rule application. The deduction rules and the logical axioms are of an easy syntactical nature and can be checked algorithmically. The same holds for the axioms of PA. We can write a computer program that tells us whether an $\mathcal{L}_A$-formula is an axiom of PA. This is crucial for our algorithm to work. If the axioms of PA were some random collection of formulas, the procedure above could not be implemented as an algorithm.

Second, since we assume that PA is complete, the algorithm will terminate for any input. By the Church-Turing thesis, $P(x)$ is computable.

The computability of $P(x)$ implies that the set

$$\ulcorner \mathsf{PA}_\vDash \urcorner = \{\ulcorner \sigma \urcorner : \mathsf{PA} \vdash \sigma\}$$

is computable. As we argued earlier in this section, if PA is complete, then $\mathsf{PA}_\vDash = \mathrm{Th}(\mathbb{N})$. Together, these two facts imply that the function g from Theorem 4.36 is computable, contradicting Theorem 4.36. □

Since for any sentence σ, either σ or $\neg\sigma$ must hold in $\mathbb{N}$, it follows that *there are true statements about $\mathbb{N}$ that PA cannot prove.*

Gödel's first incompleteness theorem actually went a lot further than the theorem above. He was able to show that *any computable, consistent extension of PA* (in fact, a very simple, finite subsystem of PA) *is incomplete*. In other words, no matter how we try to extend PA to make it into a complete theory, as long as we stay consistent and we are able to algorithmically decide whether a formula is an axiom, we are bound to fail.

Of course, the incompleteness theorem raises another question: What does a statement that is true but not provable look like? It is possible to extract such a statement from the undecidability of the halting problem and the connection between computation and definability in arithmetic. Gödel himself gave an explicit statement that is based on similar ideas.

However, these statements, as theorems of arithmetic, seem rather artificial in that they are of a self-referential nature. (Gödel's proof is often described as a formalization of the *liar paradox* regarding the impossibility of assigning a truth value to the sentence "*This statement is false.*") Are there "normal" number-theoretic theorems not provable in PA? In 1977, Paris and Harrington [**47**] were able to produce such a statement using Ramsey theory. In fact, we have already encountered this statement in Chapter 3; it is the statement we called the fast Ramsey theorem:

For all $m, p, r \geq 1$, there exists N such that for every r-coloring of $[N]^p$, there exists a homogeneous subset H of size at least m such that $|H| > \min H$.

Note that all objects in the statement are of a finite nature, and it is possible to formalize this statement in $\mathcal{L}_A$ via a process like the one we outlined for the finite Ramsey theorem in Section 4.3.

Exercise 4.39. Using the coding functions introduced in Section 4.3, formalize the statement of the fast Ramsey theorem in $\mathcal{L}_A$.

If you go back to Section 3.5 and revisit the proof of the fast Ramsey theorem (Theorem 3.40), you will see that we used compactness to infer it from the infinite Ramsey theorem. The problem with formalizing this proof in PA is that the coding methods in Section 4.3 apply only to *finite sequences* of numbers, not infinite sets. Cardinality considerations aside (there are uncountably many subsets of $\mathbb{N}$), one can devise other effective coding methods for a certain subfamily of subsets of $\mathbb{N}$ (for example, consider Gödel numbers of Turing machines computing the set). But it is possible to show that the infinite Ramsey theorem is not formalizable in PA for such an effective coding. This was proved by Jockusch [**37**].

The question is whether there might be an alternative proof of the fast Ramsey theorem that utilizes only finite objects, as it is possible for the finite Ramsey theorem. In 1977, Paris and Harrington [**47**] showed that this is impossible, and this impossibility result has become known as the **Paris-Harrington theorem**.

4.5. Indiscernibles

We will devote the remainder of this chapter to proving this result. Our presentation follows the approach of Kanamori and McAloon [38] and is greatly inspired by the account given in the book by Marker [44].

4.5. Indiscernibles

We want to show that a certain statement of first-order arithmetic is not provable in PA. How do we accomplish this? By Gödel's completeness theorem, we can show that

$$\text{PA} \nvdash \sigma$$

by constructing a model $\mathcal{M}$ of PA such that $\mathcal{M} \nvDash \sigma$. The compactness theorem in turn gave us a tool to construct models for PA other than $\mathbb{N}$, *non-standard* models. So far, however, we have had little control over the nature of a non-standard model, other than that it has non-standard elements.

In this section, we will describe a technique that allows us to construct non-standard models with additional properties. Perhaps somewhat surprisingly, Ramsey theory will play a key role.

When two objects are identical, they will have exactly the same properties. But what about the converse: If two objects have the exact same properties, are they identical?

While this is ultimately a philosophical question, there is a way to frame it mathematically. We could, for instance, say that two elements a and b of an $\mathcal{L}$-structure $\mathcal{M}$ are *indiscernible* if we cannot tell them apart by any $\mathcal{L}$-formula $\varphi(x,y)$. That is, for any such formula,

$$\mathcal{M} \vDash \varphi[a,b] \Leftrightarrow \mathcal{M} \vDash \varphi[b,a].$$

For example, we can consider a graph $G = (V, E)$ as a structure over the language $\mathcal{L} = \{E\}$ with just one binary relation symbol (the edge relation). In a complete graph K_n, any two vertices would be indiscernible in this sense.

In models of PA, however, we have a general obstruction to indiscernibility—the models are linearly ordered. For any two elements $a \neq b$, either $a < b$ or $b < a$. Therefore, the formula $\varphi(x,y) \equiv x < y$ will *discern* a from b.

We can take this into account when defining indiscernibility. This leads to the notion of *order indiscernibles*.

Definition 4.40. Let Γ be a set of $\mathcal{L}_\mathcal{A}$-formulas. We say that $X \subseteq \mathbb{N}$ is a set of **order indiscernibles for** Γ if for every formula $\varphi(x_1, \ldots, x_m) \in \Gamma$, the following holds: If φ has m free variables, then for every pair of sequences $a_1 < \cdots < a_m$ and $b_1 < \cdots < b_m$ from X,

$$\mathbb{N} \vDash \varphi[a_1, \ldots, a_m] \Leftrightarrow \mathbb{N} \vDash \varphi[b_1, \ldots, b_m].$$

There is a natural connection between formulas and colorings: Given a formula $\varphi(x_1, \ldots, x_m)$, it induces a 2-coloring of $[\mathbb{N}]^m$ by coloring $\{a_1, \ldots, a_m\}$ blue if $\mathbb{N} \vDash \varphi[a_1, \ldots, a_m]$ and red if $\mathbb{N} \nvDash \varphi[a_1, \ldots, a_m]$. We can therefore use Ramsey's theorem to construct sets of indiscernibles.

Theorem 4.41. *For every finite set of $\mathcal{L}_\mathcal{A}$-formulas Γ there exists an infinite set of order indiscernibles for Γ.*

Proof. Suppose $\Gamma = \{\varphi_1, \ldots, \varphi_k\}$. There are at most finitely many variables with a free occurrence in any formula in Γ. By re-indexing variables, we can assume that every formula in Γ has free variables among $x_1, \ldots, x_l$.

We consider the 2-coloring of $[\mathbb{N}]^l$ associated with each formula:

$$c_i(a_1, \ldots, a_l) = \begin{cases} 1 & \text{if } \mathbb{N} \vDash \varphi_i[a_1, \ldots, a_l], \\ 0 & \text{if } \mathbb{N} \nvDash \varphi_i[a_1, \ldots, a_l]. \end{cases}$$

This way, any l-element subset $\{a_1, \ldots, a_l\}$ of $\mathbb{N}$ receives k colorings, one for each formula $\varphi_1, \ldots, \varphi_k$. We can collect these colorings in one "big" coloring with 2^k different colors. The "color" of $\{a_1, \ldots, a_l\}$ in this coloring corresponds to the set

$$\{i : \mathbb{N} \vDash \varphi_i[a_1, \ldots, a_l]\}.$$

By the infinite Ramsey theorem (Theorem 2.1), there exists an infinite, homogeneous subset $H \subset \mathbb{N}$ for this coloring. By definition of the 2^k-coloring, H is a set of indiscernibles for Γ. □

Exercise 4.42. Use a compactness argument to show that if Γ is the set of *all* $\mathcal{L}_\mathcal{A}$-formulas, there exists a model of PA with an infinite set of indiscernibles for Γ.

4.5. Indiscernibles

(*Hint:* Follow the construction of a non-standard model from Section 4.2. Start with PA. Add new constant symbols $(c_i)_{i \in \mathbb{N}}$ to the language, together with new axioms forcing these constants to be pairwise distinct and indiscernible. Use compactness and the previous theorem to show that the new theory has a model.)

If we strengthen the notion a bit more, indiscernibles will help us construct models of PA. To simplify notation, we write $\vec{a} \in M$ for a tuple of elements of M and $\vec{a} < b$ to express that every entry of $\vec{a}$ is less than b.

Definition 4.43. Let Γ be a set of $\mathcal{L}_\mathcal{A}$-formulas, and suppose $\mathcal{M}$ is a model of PA. A set $X \subseteq M$ is a set of **diagonal indiscernibles** for Γ if for every $\varphi(x_1, \ldots, x_m, y_1, \ldots, y_n) \in \Gamma$, for any

$$b, b_1, \ldots, b_n, b_1^*, \ldots, b_n^* \in X$$

with $b < b_1 < \cdots < b_n$ and $b < b_1^* < \cdots < b_n^*$, and for all $\vec{a} \in M$ with $\vec{a} < b$,

$$\mathcal{M} \vDash \varphi[\vec{a}, \vec{b}] \iff \mathcal{M} \vDash \varphi[\vec{a}, \vec{b}^*]$$

Diagonal indiscernibles seem rather technical, but we will see next what the more involved definition is needed for. Let us not worry for the moment about the existence of diagonal indiscernibles, but assume we are given an infinite sequence $b_1 < b_2 < \cdots$ of diagonal indiscernibles in some model $\mathcal{M}$ of PA. How can we use this to construct another model of PA? The basic idea is to look at the *initial segment* defined by a set of diagonal indiscernibles in a model of PA,

$$N = \{y \in M : y < b_i \text{ for some } i \in \mathbb{N}\}.$$

It turns out that the indiscernibles are growing fast enough to make N closed under addition and multiplication, and are "indiscernible" enough to make it closed under induction, and thus yield a model of PA.

Before we prove this, however, we need to introduce an important type of formula.

Δ_0-formulas. The complexity of a mathematical statement can be tied to the number of quantifiers it contains or, more precisely, the number of quantifier changes. Many students find the ε-δ definition of continuity to be more difficult to handle than, say, the definition of the range of a function. Continuity is a $\forall\exists\forall$-statement, while the range is defined via a simple $\exists$-formula.

Following this idea, the simplest statements are those with no quantifiers at all. In the language of arithmetic, any $\mathcal{L}_\mathcal{A}$-sentence without quantifiers is indeed very easy, such as

$$\underline{3} + \underline{5} = \underline{8} \quad \text{or} \quad \underline{1} + \underline{1} \neq \underline{334} \lor \underline{2} = \underline{2}.$$

The left formula does not contain any *logical* symbols other than '='. Such $\mathcal{L}_\mathcal{A}$-formulas are called **atomic**—they cannot be broken up further into simpler subformulas. The formula on the right is a *Boolean combination* of atomic formulas. The atomic parts are $\varphi \equiv \underline{1} + \underline{1} = \underline{334}$ and $\psi \equiv \underline{2} = \underline{2}$, and the formula is given as $\neg\varphi \lor \psi$.

The important point about quantifier-free $\mathcal{L}_\mathcal{A}$-formulas is that, for the standard model $\mathbb{N}$, we can check whether these statements are true by means of a computer program. For statements with quantifiers, such as the Goldbach conjecture (a $\forall$-statement), this may no longer be possible. In a "brute force" attempt, a computer would have to check infinitely many instances (every even integer) and hence, if the conjecture is true, run forever.

Similarly, an $\exists$-statement can be interpreted as an *unbounded search*, since we are looking for a witness to the existential statement over the whole structure. If no such witness exists, our search will go on forever, and how and when would we decide whether that's the case? This is essentially the same question as the halting problem, which we have seen to be undecidable (Theorem 4.35).

However, if we bound our search in advance, say by looking only at numbers less than 10^6, we know that eventually our search will end, either because it has found a witness, or because there is none among the numbers up to 10^6. It might take a long time, but it will end. Syntactically, a bounded search corresponds to a *bounded quantifier*. For example,

$$\forall x < \underline{65536}\,(x > 2 \text{ even} \Rightarrow \exists p_1, p_2 < x\,(p_1, p_2 \text{ prime} \land p_1 + p_2 = x))$$

4.5. Indiscernibles

asserts that the Goldbach conjecture holds up to 2^{16} (which in fact it does). The bound in a bounded quantifier

$$\forall x < t \quad \text{or} \quad \exists x < t$$

can be any term t in which x does not occur. A term is either a constant (such as $\underline{65536}$), a variable, or an expression that can be formed from constants and variables using $+$ and $\cdot$ (such as $\underline{3} \cdot y + \underline{4}$). Formally, $(\exists x < t)\,\varphi$ and $(\forall x < t)\,\varphi$ are seen as abbreviations for $\mathcal{L}_\mathcal{A}$-formulas

$$\exists x\,(x < t \wedge \varphi) \quad \text{and} \quad \forall x\,(x < t \Rightarrow \varphi)$$

respectively.

Definition 4.44. An $\mathcal{L}_\mathcal{A}$-formula is a Δ_0**-formula** if all its quantifiers are bounded.

The intuition that a formula containing only *bounded* quantifiers is algorithmically verifiable is made rigorous by the following proposition.

Proposition 4.45. *Let* $\varphi(x_1, \ldots, x_n)$ *be a* Δ_0-*formula. Then the relation* $R \subseteq \mathbb{N}^n$ *defined by* φ,

$$R(a_1, \ldots, a_n) :\Leftrightarrow \mathbb{N} \vDash \varphi[a_1, \ldots, a_n],$$

is primitive recursive (and therefore computable).

Informally, one proves the proposition by showing that atomic formulas give rise to primitive recursive relations (as they invoke only very simple functions) and then using the closure of primitive recursive functions under the bounded μ-operator (Proposition 3.21).

While Theorem 4.36 states that in general the truth of sentences over $\mathbb{N}$ is undecidable, Proposition 4.45 establishes that if we restrict ourselves to a family of simple formulas (the Δ_0 ones), we can decide their truth value effectively—in fact, primitive recursively.

Consider the set

$$\mathrm{Sat} = \left\{ (l, c, a) : c \text{ is the Gödel number of a } \Delta_0\text{-formula } \varphi(x_1, \ldots, x_l) \right.$$

with l free variables,

a codes a sequence $a_1, \ldots, a_l$ of length l, and

$$\left. \mathbb{N} \vDash \varphi[a_1, \ldots, a_l] \right\}.$$

Since the Gödel numbering of formulas is effective, Proposition 4.45 implies that Sat is a computable set.[7] Recall from Section 4.4 that every computable set is definable (Theorem 4.37). Hence there exists a formula $\varphi_{\mathrm{Sat}}(x, y, z)$ such that $\mathbb{N} \vDash \varphi_{\mathrm{Sat}}[l, c, a]$ if and only if c is the Gödel number of a Δ_0-formula $\varphi(x_1, \ldots, x_l)$ with l free variables, a codes a sequence $a_1, \ldots, a_l$ of length l, and $\mathbb{N} \vDash \varphi[a_1, \ldots, a_l]$. So we can capture the truth of Δ_0-formulas by a formula. Even better, this formula is again rather simple (not quite Δ_0, but close), so simple that it works even in other models of PA. This will be tremendously useful later on, and we will come back to it in due time.

Models from indiscernibles. We now put the Δ_0-formulas to use in finding new models of PA inside given models.

Proposition 4.46. *Suppose that $\mathcal{M}$ is a model of Peano arithmetic and $b_1 < b_2 < \cdots$ is a sequence of diagonal indiscernibles for all Δ_0-formulas. Let N be the initial segment of M given by*

$$N = \{ y \in M : y < b_i \text{ for some } i \in \mathbb{N} \}.$$

Then N is closed under S, $+$, and $\cdot$, and hence $\mathcal{M}$ restricted to N forms an $\mathcal{L}_\mathcal{A}$-structure, denoted by $\mathcal{N}$. Furthermore, $\mathcal{N}$ is a model of PA.

Proof.

(i) *Closure under basic arithmetic operations:* Closure under S is easy, since in any model of PA it holds that if $a < b$ and $b < c$, then $S(a) < c$ (this can be proved via induction). And so, as we have an infinite strictly increasing sequence $b_1 < b_2 < \cdots$, N is closed under S.

---

[7] Sat stands for *satisfaction relation* ($\vDash$).

4.5. Indiscernibles

Addition is a bit harder. Here we need to use indiscernibility for the first time. Suppose $x < b_i$ and $y < b_j$. We need to show that $x + y < b_k$ for some k. It turns out that we can take any b_k with $k > i, j$ (which tells us that the b_i must be growing rather quickly). For suppose $x + y \geq b_k$. As $y < b_j$, this means $x + b_j > b_k$. As $\mathcal{M}$ satisfies all the rules of elementary arithmetic, we can find $a \in M$ with $a < x$ and $a + b_j = b_k$. Now indiscernibility strikes: Consider the formula

$$\varphi(x, y, z) \equiv x + y = z.$$

$\mathcal{M} \vDash \varphi[a, b_j, b_k]$ holds and hence by indiscernibility $\mathcal{M} \vDash \varphi[a, b_j, b_l]$ for any $b_l > b_k$ (you should check that all the necessary assumptions for diagonal indiscernibility are met), which, by elementary arithmetic, would imply that

$$b_k = b_l,$$

a contradiction. So $x + y < b_k$, as desired.

We also want to deduce $xy < b_k$ for $k > i, j$, thereby showing that N is closed under multiplication. Suppose $xy \geq b_k$. This would imply $xb_j > b_k$. We cannot divide in models of PA, so we cannot assume the existence of an $a < x$ such that $ab_j = b_k$. Instead, we pick a such that

(4.7) $$ab_j < b_k \leq (a+1)b_j.$$

Applying indiscernibility, we get $ab_j < b_l \leq (a+1)b_j$ for any $l > k$. Now add b_j to all parts of (4.7), and we get

$$(a+1)b_j < b_k + b_j \leq (a+2)b_j,$$

and thus

$$b_l \leq (a+1)b_j < b_j + b_k.$$

But the argument for addition yielded that $b_j + b_k < b_l$ whenever $j, k < l$, so we have arrived at a contradiction and must conclude that $xy < b_k$ for any $k > i, j$.

(ii) $\mathcal{N}$ *satisfies the axioms of* PA: $\mathcal{N}$ inherits the properties of the bigger structure $\mathcal{M}$ to a certain extent. The axioms (PA1) through (PA6) state that a simple property (such as $x + 0 = x$) holds *for all* elements. In fact, all of the axioms (PA1) through (PA6) are of the form

$$\forall \vec{x} \, [\Delta_0\text{-formula}].$$

The truth values of Δ_0-formulas themselves persist when we pass to an initial segment of $\mathcal{M}$ closed under the basic arithmetic operations (such as our N here).

Lemma 4.47. *Suppose* $\mathcal{M}$ *is a model of* PA *and* $N \subseteq M$ *is such that*

(1) N *is closed under* $S, +, \cdot$ *(as given by* $\mathcal{M}$*), and*

(2) N *is an initial segment, that is, if* $b \in N$ *and* $a \in M$ *is such that* $a < b$*, then* $a \in N$.

(These two conditions in particular imply that the structure $\mathcal{N} = (N, S^{\mathcal{M}} \restriction_N, +^{\mathcal{M}} \restriction_N, \cdot^{\mathcal{M}} \restriction_N, 0^{\mathcal{M}})$ *is an* $\mathcal{L}_{\mathcal{A}}$*-structure.) Then for any* Δ_0*-formula* $\varphi(x_1, \ldots, x_n)$ *and any* $a_1, \ldots, a_n \in N$,

$$\mathcal{N} \vDash \varphi[a_1, \ldots, a_n] \iff \mathcal{M} \vDash \varphi[a_1, \ldots, a_n].$$

Proof sketch. The lemma is proved by *induction on the formula structure*. This means that we first verify the claim for *atomic* formulas (in the case of $\mathcal{L}_{\mathcal{A}}$-formulas these are just equations over $S, +, \cdot$ and 0). Then, assuming we have verified the claim for formulas φ and ψ, we deduce it for formulas $\neg \varphi$, $\varphi \vee \psi$, and $(\exists x < t)\psi$. Most cases are straightforward, since atomic relations are not affected when passing to a smaller structure. Only quantification is an issue, as the witness to an $\exists$-statement in $\mathcal{M}$ may no longer be available in $\mathcal{N}$. But in a *bounded* statement, these witnesses have to come from the smaller structure, because it is an initial segment.

A full proof has to address some technical issues such as the interpretation of terms, which we have avoided in our presentation in order to keep things simple. You can find a full proof in most textbooks on mathematical logic, such as [54]. □

Equipped with this lemma, we can argue that axioms (PA1)–(PA6) hold in $\mathcal{N}$. In fact, we argue in general that for any sentence φ of the form

$$\forall \vec{x}\, \psi(\vec{x}),$$

where ψ is a Δ_0-formula, we have that

$$\mathcal{M} \vDash \varphi \quad \text{implies} \quad \mathcal{N} \vDash \varphi.$$

4.5. Indiscernibles

For suppose $\mathcal{M} \models \varphi$. This means that *for all $\vec{a} \in M$, $\mathcal{M} \models \psi[\vec{a}]$*. In particular, $\mathcal{M} \models \psi[\vec{a}]$ for all $\vec{a} \in N$. By Lemma 4.47, $\mathcal{N} \models \psi[\vec{a}]$ for all $\vec{a} \in N$, which in turn implies that $\mathcal{N} \models \psi$.

It might be tempting to apply the same argument to the induction scheme (Ind_φ), as on the surface it looks as if there are only universal quantifiers ($\forall$) occurring. But keep in mind that φ can be *any* formula, and for more complicated formulas it is not clear at all how the properties they define would persist when passing from $\mathcal{M}$ to $\mathcal{N}$. This is where indiscernibles come in: They allow us to reduce truth in $\mathcal{N}$ to truth for Δ_0-formulas, and as we saw above, Δ_0-formulas are very well behaved when passing from $\mathcal{M}$ to a smaller substructure. Recall that the induction axiom for φ is

$$(\text{Ind}_\varphi) \quad \forall \vec{w}[(\varphi(0, \vec{w}) \wedge \forall v\,(\varphi(v, \vec{w}) \to \varphi(S(v), \vec{w}))) \to \forall v\, \varphi(v, \vec{w})].$$

We first simplify our life a bit by only considering formulas of a certain canonical form. A formula is in *prenex normal form* if all quantifiers are on the "outside". For example, if ψ is quantifier free, then

$$(4.8) \qquad \exists x_1 \forall x_2 \ldots \exists x_n\, \psi(v, \vec{w}, \vec{x})$$

is in prenex normal form. Every formula is logically equivalent to a formula in prenex normal form via logical equivalences such as

$$(\exists x\, \varphi) \vee \psi \text{ is equivalent to } \exists x(\varphi \vee \psi),$$

provided x does not occur as a free variable in ψ, and

$$\neg \exists x \varphi \text{ is equivalent to } \forall x \neg \varphi.$$

Hence we can assume that φ is in prenex normal form. Moreover, observe that in the above example, $\exists$ and $\forall$ are alternating. We can also assume this for φ, since using coding we can contract multiple quantifiers to one. For example, in PA,

$$\exists x_1 \exists x_2 \varphi(x_1, x_2)$$

is equivalent to

$$\exists x \varphi(\text{decode}(x, 1), \text{decode}(x, 2)).$$

So, by contracting quantifiers and possibly adding "dummy" variables and expressions like $x_i = x_i$, we can assume that φ is of the form given in (4.8).

Let us consider first a formula $\exists x\, \theta(x)$. Since the b_i are unbounded in N, we have

$\mathcal{N} \vDash \exists x\, \theta(x)$ iff   exists $a \in N$ such that $\mathcal{N} \vDash \theta[a]$,
                     iff   exist $a \in N, i \in \mathbb{N}$ such that $a < b_i$ and $\mathcal{N} \vDash \theta[a]$,
                     iff   exists $i \in \mathbb{N}$ such that $\mathcal{N} \vDash (\exists x < y\, \theta(x))[b_i]$.

While the b_i are, strictly speaking, not part of the language, for the sake of readability, in what follows we write the latter expression as

$$\mathcal{N} \vDash \exists x < b_i\, \theta(x).$$

Similarly, we have

$\mathcal{N} \vDash \forall x\, \theta(x)$ iff   for all $i \in \mathbb{N}$, $\mathcal{N} \vDash \forall x < b_i\, \theta(x)$.

Now suppose $\varphi(v, \vec{w})$ written in prenex normal form is

$$\exists x_1 \forall x_2 \ldots \exists x_n\, \psi(v, \vec{w}, \vec{x})$$

and $a, \vec{c} \in N$. Choose i_0 such that $a, \vec{c} < b_{i_0}$. Inductively applying the equivalences above, we get that[8]

$$\mathcal{N} \vDash \exists x_1 \forall x_2 \ldots \exists x_n\, \psi(a, \vec{c}, \vec{x})$$

is equivalent to

$\exists i_1 > i_0\, \forall i_2 > i_1\, \ldots\, \exists i_n > i_{n-1}$
$$\mathcal{N} \vDash \exists x_1 < b_{i_1}\, \forall x_2 < b_{i_2}\, \ldots\, \exists x_n < b_{i_n}\, \psi(a, \vec{c}, \vec{x}).$$

Note that the "meta"-quantifiers, $\exists i_1 > i_0\, \forall i_2 > i_1\, \ldots\, \exists i_n > i_{n-1}$, are just an abbreviation of the long statement "there exists $i_1 > i_0$ such that for all ...". The formula on the right-hand side is Δ_0, since all quantified variables are bounded, and hence by Lemma 4.47 the previous statement is equivalent to

$\exists i_1 > i_0\, \forall i_2 > i_1\, \ldots\, \exists i_n > i_{n-1}$
$$\mathcal{M} \vDash \exists x_1 < b_{i_1}\, \forall x_2 < b_{i_2}\, \ldots\, \exists x_n < b_{i_n}\, \psi(a, \vec{c}, \vec{x}).$$

---

[8] The notation in the succeeding formula is again a little sloppy. As a and $\vec{c}$ are not variables but elements of the structure over which we interpret, we should write $\psi(\vec{x})[a, \vec{c}]$, but that makes it rather hard to read.

4.5. Indiscernibles

Since the b_i are diagonal indiscernibles for all Δ_0-formulas, it does not really matter which b_j we use to bound the quantifiers, so the last statement is equivalent to

$$\mathcal{M} \vDash \exists x_1 < b_{i_0+1} \, \forall x_2 < b_{i_0+2} \ldots \exists x_n < b_{i_0+n} \, \psi(a, \vec{c}, \vec{x}).$$

To sum up, if $a, \vec{c} \in N$ and i_0 is such that $a, \vec{c} < b_{i_0}$, then

$$\mathcal{N} \vDash \varphi[a, \vec{c}] \text{ iff } \mathcal{M} \vDash \exists x_1 < b_{i_0+1} \, \forall x_2 < b_{i_0+2} \ldots \exists x_n < b_{i_0+n} \, \psi(a, \vec{c}, \vec{x}).$$

Note that the statement on the right-hand side is a "pure" Δ_0-statement; it has no more meta-quantifiers.

We finally show that $\mathcal{N}$ satisfies induction. Recall that (Ind) is equivalent to the *least number principle* (LNP); see Section 4.1. Suppose $\mathcal{N} \vDash \varphi[a, \vec{c}]$, where $\varphi(v, \vec{w})$ is given in prenex normal form as

$$\exists x_1 \forall x_2 \ldots \exists x_n \, \psi(v, \vec{w}, \vec{x}), \quad \text{with } \psi \text{ quantifier free.}$$

As before, we choose i_0 such that $a, \vec{c} < b_{i_0}$ and obtain the equivalence

$$\mathcal{N} \vDash \varphi[a, \vec{c}] \text{ iff } \mathcal{M} \vDash \exists x_1 < b_{i_0+1} \, \forall x_2 < b_{i_0+2} \ldots \exists x_n < b_{i_0+n} \, \psi(a, \vec{c}, \vec{x}).$$

Since induction (and hence the LNP) holds in $\mathcal{M}$, there exists a *least* $\hat{a} < b_{i_0}$ such that

$$\mathcal{M} \vDash \exists x_1 < b_{i_0+1} \, \forall x_2 < b_{i_0+2} \ldots \exists x_n < b_{i_0+n} \, \psi(\hat{a}, \vec{c}, \vec{x}).$$

By the definition of $\mathcal{N}$, the existence of $\hat{a} \in N$, and the equivalence above, $\mathcal{N} \vDash \varphi[\hat{a}, \vec{c}]$. Finally, $\hat{a}$ has to be the smallest witness to φ in $\mathcal{N}$, because any smaller witness would also be a smaller witness in $\mathcal{M}$. This concludes the proof of Proposition 4.46. □

The construction of $\mathcal{N}$ hinged on two crucial points:

- the use of indiscernibles to relate truth of general formulas to truth of Δ_0-formulas, and
- the *absoluteness* (that is, the persistence of truth) of Δ_0-formulas between models of PA and sufficiently closed initial segments.

Our next goal is to actually construct a sequence of diagonal indiscernibles. For this, we return to Ramsey theory.

4.6. Diagonal indiscernibles via Ramsey theory

We have already seen how to use Ramsey's theorem to construct indiscernibles. Can we use the same technique to construct *diagonal* indiscernibles? The key idea in using Ramsey theory in this context was to *color* sets $\{a_1, \ldots, a_l\}$ according to their truth behavior in $\mathbb{N}$ with respect to formulas φ_i. A homogeneous subset then guaranteed that all l-subsets from the set have the same truth behavior.

For diagonal indiscernibles, however, the situation is more complicated. If you revisit Definition 4.43, you will see that the notion depends on the truth behavior not only among the indiscernibles themselves, but also in relation to arbitrary numbers that are all dominated by the tuple of indiscernibles in question. This will require a slightly different notion of homogeneity, which is, as we will see, closely related to the fast Ramsey theorem (Theorem 3.40).

Definition 4.48. Let $X \subseteq \mathbb{N}$ and $n \geq 1$, and suppose $f : [X]^n \to \mathbb{N}$. A set $M \subseteq X$ is **min-homogeneous** if for f and only $s, t \in [M]^n$,

$$\min s = \min t \Rightarrow f(s) = f(t).$$

A function $f = (f_1, \ldots, f_m) : [X]^n \to \mathbb{N}^m$ is min-homogeneous if every component f_i is.

On a min-homogeneous set, f depends only on the least argument, just as the truth behavior of diagonal indiscernibles only depends on the lower bound of a tuple of indiscernibles.

We now show how to use min-homogeneous functions to construct diagonal indiscernibles in $\mathbb{N}$.

4.6. Diagonal indiscernibles via Ramsey theory

Lemma 4.49. *For any $c, e, k, n, m \geq 1$ and Δ_0-formulas*
$$\varphi_1(x_1, \ldots, x_m, y_1, \ldots, y_n), \ldots, \varphi_e(x_1, \ldots, x_m, y_1, \ldots, y_n),$$
there is a k-element subset H of $\mathbb{N}$ such that H is a set of diagonal indiscernibles for $\varphi_1, \ldots, \varphi_e$ and $\min H \geq c$.

Proof. For technical purposes, we assume $k > 2n$.

By the finite Ramsey theorem, there exists a $w \in \mathbb{N}$ such that
$$w \longrightarrow (n+k)_{e+1}^{2n+1}.$$
Now let us assume that W is a sufficiently large number (how large we will see below). We define two functions, $f = (f_1, \ldots, f_m)$ and i, on $(2n+1)$-element subsets of $\{0, \ldots, W-1\}$. Let $b_0 < b_1 < \cdots < b_{2n} < W$. If for all $\vec{a} < b_0$ and all j,
$$\mathbb{N} \vDash \varphi_j[\vec{a}, b_1, \ldots, b_n] \iff \mathbb{N} \vDash \varphi_j[\vec{a}, b_{n+1}, \ldots, b_{2n}],$$
then let
$$f(b_0, b_1, \ldots, b_{2n}) = (0, \ldots, 0) \text{ and } i(b_0, b_1, \ldots, b_{2n}) = 0.$$
If there exist $\vec{a} < b_0$ and j such that
$$\mathbb{N} \vDash \varphi_j[\vec{a}, b_1, \ldots, b_n] \not\iff \mathbb{N} \vDash \varphi_j[\vec{a}, b_{n+1}, \ldots, b_{2n}],$$
fix any such $\vec{a}$ and j and put
$$f(b_0, b_1, \ldots, b_{2n}) = \vec{a} \text{ and } i(b_0, b_1, \ldots, b_{2n}) = j.$$

Now suppose W is so large that there exists a subset $H_0 \subseteq \{0, \ldots, W-1\}$ such that H_0 is min-homogeneous for f and
$$|H_0| \geq w \text{ and } \min H_0 \geq c.$$
It is not obvious at all that such a W exists, but let us assume for now that it does exist. We will of course have to return to this point later (Lemma 4.52).

We continue by homogenizing H_0 with respect to i: By choice of w, there exists a subset $H_1 \subseteq H_0$ of cardinality $k + n$ such that i is constant on $[H_1]^{2n+1}$. The ideal case for us is that the homogeneous value is $i \equiv 0$. This means that all the $(2n+1)$-sets B behave the same with respect to all formulas φ_j, and we can use this to define a set of diagonal indiscernibles: Let $H = \{b_1 < \cdots < b_k\}$ be the first k elements

of H_1. Since $|H_1| = k+n$, H_1 has n additional elements, which we call $z_1 < \cdots < z_n$. Thus
$$b_1 < \cdots < b_k < z_1 < \cdots < z_n.$$
Now for any $b_0 < b_1 < \cdots < b_n$ and $b_0 < b_1^* < \cdots < b_n^*$ from H, since $f(b_0, b_1, \ldots, b_n, z_1, \ldots, z_n) = \vec{0}$ and $f(b_0, b_1^*, \ldots, b_n^*, z_1, \ldots, z_n) = \vec{0}$, we have that
$$\mathbb{N} \vDash \varphi_j[\vec{a}, b_1, \ldots, b_n] \Leftrightarrow \mathbb{N} \vDash \varphi_j[\vec{a}, z_1, \ldots, z_n] \Leftrightarrow \mathbb{N} \vDash \varphi_j[\vec{a}, b_1^*, \ldots, b_n^*]$$
for all $\vec{a} < b_0$.

What about the other cases? Assume $i \equiv j > 0$. Let $b_0 < \cdots < b_{3n}$ be elements from H_1. This is where we use the assumption $k \geq 2n+1$. By the min-homogeneity of f on H_1, there exists $\vec{a} < b_0$ such that
$$f(b_0, b_1, \ldots, b_n, b_{n+1}, \ldots, b_{2n}) = f(b_0, b_1, \ldots, b_n, b_{2n+1}, \ldots, b_{3n})$$
$$= f(b_0, b_{n+1}, \ldots, b_{2n}, b_{2n+1}, \ldots, b_{3n})$$
$$= \vec{a}.$$
This means that
$$\mathbb{N} \vDash \varphi_j[\vec{a}, b_1, \ldots, b_n] \not\Leftrightarrow \mathbb{N} \vDash \varphi_j[\vec{a}, b_{n+1}, \ldots, b_{2n}],$$
$$\mathbb{N} \vDash \varphi_j[\vec{a}, b_1, \ldots, b_n] \not\Leftrightarrow \mathbb{N} \vDash \varphi_j[\vec{a}, b_{2n+1}, \ldots, b_{3n}],$$
$$\mathbb{N} \vDash \varphi_j[\vec{a}, b_{n+1}, \ldots, b_{2n}] \not\Leftrightarrow \mathbb{N} \vDash \varphi_j[\vec{a}, b_{2n+1}, \ldots, b_{3n}].$$
But this is impossible, since there are only two possible truth values. Therefore, the case $i > 0$ is impossible, and we have constructed a set of diagonal indiscernibles for $\varphi_1, \ldots, \varphi_e$. □

We took a leap of faith in the above proof and assumed that we can always find min-homogeneous functions, provided our choice of W is large enough. The problem here is that we do not have a coloring by a fixed number of colors in the traditional sense. The values of f can be arbitrarily large. They are bounded by $W - 2n - 1$, but W is the number we are trying to find, so if we make W larger we also make the range of f larger, which in turn makes it harder to find homogeneous objects. And in general, finding even small min-homogeneous subsets can be impossible.

Exercise 4.50. Find a function $f : [\mathbb{N}]^2 \to \mathbb{N}$ for which there does not exist *any* min-homogeneous set.

4.6. Diagonal indiscernibles via Ramsey theory

However, if you look back at the function f we defined in the proof above, you should notice that for all $X \in [\mathbb{N}]^{2n+1}$, all components of $f(X)$ are smaller than $\min X$.

Definition 4.51. For $n, m \geq 1$ and $A \subseteq \mathbb{N}$, a function $f : [A]^n \to \mathbb{N}$ is called **regressive** if for all $X \in [A]^n$, $f(X) < \min X$. A vector-valued function $f = (f_1, \ldots, f_m)$ is regressive if every component f_j is.

The restriction on the range that regressiveness imposes makes it possible to find min-homogeneous sets. To do this, we use the fast Ramsey theorem.

Lemma 4.52 (Principle $(*)$). *For any $c, k, n, m \geq 1$ there exists a $W \in \mathbb{N}$ such that every regressive function $f : [W]^n \to \mathbb{N}^m$ has a min-homogeneous set $H \subseteq \{0, \ldots, W-1\}$ with $|H| \geq k$ and $\min H \geq c$.*

Proof. A slight modification of the proof of the fast Ramsey theorem yields that there exists a number W such that for every function $g : [W]^{n+1} \to 3^m$, there exists a set $Y \subseteq \{0, \ldots, W\}$ such that Y is homogeneous for g and

(4.9) $\quad |Y| \geq k + n, \quad |Y| \geq \min Y + n + 1, \text{ and } \min Y \geq c.$

(See Exercise 4.53 below.)

We claim that W is large enough to guarantee the existence of the min-homogeneous set we are looking for.

Let $f : [W]^n \to \mathbb{N}^m$ be regressive. We can write f as a vector of m functions,

$$f = (f_1, \ldots, f_m).$$

For any $b_0 < b_1 < \cdots < b_n < W$ and every $i = 1, \ldots, m$, we define

$$g_i(b_0, b_1, \ldots, b_n) = \begin{cases} 0 & \text{if } f_i(b_0, b_1, \ldots, b_{n-1}) = f_i(b_0, b_2, \ldots, b_n), \\ 1 & \text{if } f_i(b_0, b_1, \ldots, b_{n-1}) > f_i(b_0, b_2, \ldots, b_n), \\ 2 & \text{if } f_i(b_0, b_1, \ldots, b_{n-1}) < f_i(b_0, b_2, \ldots, b_n). \end{cases}$$

Together, the g_i give us a coloring

$$g = (g_1, \ldots, g_m) : [W]^{n+1} \to \{0, 1, 2\}^m.$$

For the purpose of this proof, we can identify g with a coloring $g : [W]^{n+1} \to 3^m$.

By the choice of W, there exists a set Y that is homogeneous for g and satisfies (4.9). As g is constant on $[Y]^{n+1}$, so are all the g_i. We claim that the value of each g_i on $[W]^{n+1}$ is 0. This will let us find a min-homogeneous subset (for f) in W. The idea is that if $g_i|_{[W]^{n+1}} \not\equiv 0$, f_i would have to take too many values for f to be regressive.

Fix $i \leq m$ and let us first list Y as

$$Y = \{y_0 < y_1 < \cdots < y_s\}.$$

If we let $Y_j = \{y_j, y_{j+1}, \ldots, y_{j+n-1}\}$ for $j = 1, \ldots, s - n + 1$, then, since f is regressive,

$$f_i(\{y_0\} \cup Y_j) < y_0 \text{ for all } j.$$

How many different Y_j are there? This is where we need the bounds on $|Y|$:

$$s + 1 = |Y| \geq \min Y + n + 1 = y_0 + n + 1,$$

and hence

$$s - n + 1 \geq y_0 + 1.$$

Thus, there are at least $y_0 + 1$ "blocks" Y_j for y_0 possible values $0, \ldots, y_0 - 1$, which by the pigeonhole principle means that

$$f_i(\{y_0\} \cup Y_j) = f_i(\{y_0\} \cup Y_l)$$

for some $j \neq l$. Since g_i is constant on $[Y]^{n+1}$, the sequence

$$f_i(\{y_0\} \cup Y_1), f_i(\{y_0\} \cup Y_2), \ldots, f_i(\{y_0\} \cup Y_{s-n+1})$$

is either strictly increasing, strictly decreasing, or constant. As two values are identical, the sequence must be constant and $g_i|_{[W]^{n+1}} \equiv 0$.

We can now construct our min-homogeneous set. This is similar to the proof of Lemma 4.49 in that we use some elements of Y as "anchor points" to establish min-homogeneity. Let $H = \{b_1 < \cdots < b_k\}$ be the first k elements of Y, and let $z_1 < \cdots < z_{n-1}$ be the next $n - 1$ elements of Y, that is,

$$b_1 < \cdots < b_k < z_1, \ldots, z_{n-1}.$$

4.6. Diagonal indiscernibles via Ramsey theory

As $|Y| \geq k + n$, this is possible. Suppose $b_1 < b_2 < \cdots < b_n \in H$ and also $b_1 < b_2^* < \cdots < b_n^* \in H$. Then, as $g_i|_{[W]^{n+1}} \equiv 0$ for all $i \leq m$,

$$f_i(b_1, b_2, \ldots, b_n) = f_i(b_1, b_3, \ldots, b_n, z_1)$$
$$= f_i(b_1, b_4, \ldots, b_n, z_1, z_2)$$
$$\vdots$$
$$= f_i(b_1, z_1, z_2, \ldots, z_{n-1}).$$

And likewise,

$$f_i(b_1, b_2^*, \ldots, b_n^*) = f_i(b_1, b_3^*, \ldots, b_n^*, z_1)$$
$$= f_i(b_1, b_4^*, \ldots, b_n^*, z_1, z_2)$$
$$\vdots$$
$$= f_i(b_1, z_1, z_2, \ldots, z_{n-1}).$$

Hence

$$f_i(b_1, b_2, \ldots, b_n) = f_i(b_1, b_2^*, \ldots, b_n^*),$$

and since this holds for any $i \leq m$, it shows that H is min-homogeneous for f. Finally, $\min H \geq c$ since $\min Y \geq c$. □

Exercise 4.53. Prove the modification of the fast Ramsey theorem needed in the proof above. Show that for any $k, n, r, c \in \mathbb{N}$ there exists a number W such that for every function $g : [W]^{n+1} \to \{1, \ldots, r\}$, there exists a set $Y \subseteq \{0, \ldots, W\}$ such that Y is homogeneous for g and

(4.10) $|Y| \geq k + n$, $|Y| \geq \min Y + n + 1$,   and $\min Y \geq c$.

Before we move on, it is important to reflect on the formal aspects of the previous two proofs. While we are trying to show that the fast Ramsey theorem is not provable in PA, its statement is formalizable in PA (Exercise 4.39). The same holds for Principle (∗).

If you peruse the proof of Lemma 4.52, you will notice that the steps themselves are finite combinatorial assertions in $\mathbb{N}$ that can be formalized in PA. The lemma establishes that

PA ⊢ (fast Ramsey theorem ⇒ Principle (∗)).

Exercise 4.54. Using the encoding of predicates defined in Section 4.3 (such as length(x), set(x), or decode(x, i)), formalize in $\mathcal{L}_A$ the property "the set H is an initial segment of the set Y of size n".

Lemma 4.49 is formalizable in PA, too, although this is a little harder to see. First of all, the concept of (diagonal) indiscernibles is a *semantic* notion: It is defined based on the truth behavior of formulas in a structure. If we want to formalize this in PA, we have to replace this semantic notion by something that is expressible in $\mathcal{L}_\mathcal{A}$. This is where the truth predicate φ_{Sat} comes in. It lets us talk about truth in $\mathbb{N}$ in $\mathcal{L}_\mathcal{A}$.

Moreover, to work in PA, we will have to replace the standard model $\mathbb{N}$ by an arbitrary PA-model $\mathcal{M}$. For this, we need that φ_{Sat} works *uniformly* for any model of PA.

In Lemma 4.47, we saw that truth values of Δ_0-formulas persist between models of PA and sufficiently closed initial segments. Similarly, simple formulas define properties that persist between models of PA. In $\mathcal{M}$, the formula φ_{Sat} will hold for all triples $(l, c, a) \in M^3$ such that if c is a Gödel number of $\psi(x_1, \ldots, x_l)$ and a codes $a_1, \ldots, a_l$, then $\mathcal{M} \vDash \psi[a_1, \ldots, a_l]$. You can find a thorough development of such a formula φ_{Sat} in the book by Kaye [39].

With the help of φ_{Sat}, we can formally define Δ_0-indiscernibles in $\mathcal{L}_\mathcal{A}$ and also formalize the steps of the proof.

4.7. The Paris-Harrington theorem

We can finally put all the pieces together.

Theorem 4.55 (Paris and Harrington [47]). *The fast Ramsey theorem is not provable in* PA.

We present the main steps of the proof (using the Kanamori-McAloon approach [38]) along with some important details.

Proof of the Paris-Harrington theorem.

(1) *Reduce to Principle* (∗).
The proof below will actually show that Principle (∗) (Lemma 4.52) is not provable in PA. We argued at the end of the previous section that

$$\text{PA} \vdash \text{fast Ramsey theorem} \Rightarrow (*)$$

4.7. The Paris-Harrington theorem

Hence if PA proved the fast Ramsey theorem, PA would also prove (∗), since we could first use the proof of the fast Ramsey theorem and then "append" a proof of (fast Ramsey ⇒ (∗)) to get a proof of (∗) in PA.

(2) *Use a non-standard model.*
By Lemma 4.10, to show that PA ⊬ (∗) it suffices to find a model $\mathcal{N}$ of PA in which (∗) does not hold. Using infinitary methods, we already showed that the fast Ramsey theorem holds in the standard model $\mathbb{N}$, and hence (∗) also holds in $\mathbb{N}$. So we have to look among the non-standard models of PA. Let $\mathcal{M}$ be such a model. If we happened to pick one where $\mathcal{M} \not\models (∗)$, then great, we are done. So let us assume $\mathcal{M} \models (∗)$ and let $c \in M \smallsetminus \mathbb{N}$ be a non-standard element of $\mathcal{M}$.

At this point it is important to keep the following in mind: To us, who know that $\mathcal{M}$ is a non-standard model, c "looks" infinite because it is bigger than every natural number. $\mathcal{M}$, however, "thinks" it is a perfectly normal model of PA and that c is a perfectly good citizen that looks just like every other number. In particular, as long as we are working with statements provable in PA, $\mathcal{M}$ can apply these statements to any of its elements, c or any other number. Making the distinction between "inside $\mathcal{M}$" and "outside $\mathcal{M}$" will be crucial later on.

(3) *Since the model $\mathcal{M}$ satisfies (∗), we can use it to find diagonal indiscernibles.*
This is the metamathematically most subtle step. The finite Ramsey theorem is provable in PA; there exists a least $w \in M$ such that

$$\mathcal{M} \models w \to (3c+1)_c^{2c+1}.$$

Keep in mind that w and numbers like $3c + 1$ are non-standard, too.

Since $\mathcal{M} \models (∗)$, there also exists a least $W \in M$ such that for any regressive function $f : [W]^{2c+1} \to \mathbb{N}$, there exists a set $H \subseteq \{a \in M : a < W\}$ with $\min H \geq c$ and $|H| \geq w$ such that H is min-homogeneous for f.

We should reflect for a moment what a statement such as the above means if non-standard numbers are involved. To $\mathcal{M}$ itself it looks perfectly normal: It is just an instance of the finite Ramsey theorem after all, and as we pointed out above, to $\mathcal{M}$, all its elements "look" finite. But what does it mean "from the outside" that, for example, $|H| \geq w$ if w is non-standard?

Going back to Section 4.3, the formalized version of "$|H| \geq w$" states that a code a (for H) exists such that a codes a set and the length of a is at least w. We defined the length of a code simply as the 0-entry in its decoding sequence. But can such an entry be non-standard? In other words, can the result of

$$\mathrm{rem}(c, 1 + (1+i)d)$$

be non-standard? Of course it can. If $c = d + e$, where $e < d$, then $\mathrm{rem}(c, d) = e$. This holds for *any* c, d, and e, no matter whether they are standard or not. Now we reason that the basic arithmetic facts (such as Euclidean division) are provable in PA and hence must hold in any model. And since the β-function is defined using only elementary arithmetic operations, it works in arbitrary models of PA. This means that the same formula that defines "$|H| \geq w$" in $\mathbb{N}$ (for finite sets) works also in $\mathcal{M}$, where it can code sets that look finite inside the model but infinite from the outside (recall that the operation $\mathrm{rem}(c, 1 + (1+i)d)$ is defined for *any* c, d, and i, standard or not).

We would like to find a sequence of diagonal indiscernibles in $\mathcal{M}$ and apply Proposition 4.46 to find another model of PA *inside* $\mathcal{M}$. Lemma 4.49 tells us that we can use Principle $(*)$ to find such indiscernibles. However, there is a crucial discrepancy: Proposition 4.46 requires a set of indiscernibles for *all* Δ_0-formulas, while Lemma 4.49 will give them for a *finite number* of formulas only.

Let us rephrase Lemma 4.49:

For any e, m, n, k, and l, there exists a set H with at least k elements, with minimal element $\geq l$, which is a set of diagonal indiscernibles for the first e Δ_0-formulas with at most $m + n$ free variables (according to our Gödel numbering of $\mathcal{L}_A$-formulas).

As noted in Section 4.6, this is provable from PA $+ (*)$. In particular, it holds in $\mathcal{M}$ and we can "plug in" any numbers we like, standard or

4.7. The Paris-Harrington theorem 191

non-standard. So let us do it for $k = l = c$, where c is the non-standard element chosen above.

Since we can effectively recover a formula from its Gödel number, all computable relations are definable by a relatively simple formula, and since relatively simple formulas define properties that persist between models of PA, there exists an $\mathcal{L}_A$-formula $\theta(x)$ such that for all $j \in \mathbb{N}$, $\mathcal{M} \vDash \theta[j]$ if and only if the above statement holds for $e = m = n = j$ and $k = l = c$ in $\mathcal{M}$. Clearly, $\mathcal{M} \vDash \theta[j]$ for all $j \in \mathbb{N}$. By overspill (Corollary 4.19), there exists a non-standard $b \in M$ such that $\mathcal{M} \vDash \theta[b]$.

What does $\theta[b]$ mean when b is non-standard? The statement in this case says we have diagonal indiscernibles for the first b Δ_0-formulas with at most $2b$ free variables. Non-standard Gödel numbers might code objects that are not formulas in the strict syntactical sense, but only "look" to $\mathcal{M}$ like valid Gödel numbers. The important thing, however, is that the Δ_0-formulas with Gödel number at most b include *all formulas with finite Gödel number*, and "with at most $2b$ free variables" includes *all formulas with any finite number of variables*. Furthermore, finite Gödel numbers (the ones of standard formulas) represent the same formulas between $\mathbb{N}$ and $\mathcal{M}$. This follows from the fact that being a Gödel number of an $\mathcal{L}_A$-formula is a computable, hence simply definable property and simply definable properties do not change between models of PA.

It follows—and this is a crucial fact—that H is a set of indiscernibles (over $\mathcal{M}$) for *all standard Δ_0-formulas*, and we can apply Proposition 4.46.

(4) *Use the indiscernibles to find a new PA-model inside $\mathcal{M}$.*
Now that we have a set H of at least c-many indiscernibles for all Δ_0-formulas, we can take the first $\mathbb{N}$ ones, $H' = \{h_1 < h_2 < \cdots\}$, and apply Proposition 4.46. This yields a model $\mathcal{N}$ of PA such that the universe N of $\mathcal{N}$ contains c (since all elements of the set H are $\geq c$) and does *not* contain W (because of the way the indiscernibles are constructed from a regressive function).

(5) *Because $\mathcal{N}$ is a model of PA, we can do finite Ramsey theory inside it.* If $\mathcal{N}$ does not satisfy $(*)$, we are done. So let us assume

$\mathcal{N} \vDash (*)$ and derive a contradiction. Since $\mathcal{N} \vDash \mathsf{PA}$ and PA proves the finite Ramsey theorem, and since $c \in N$, there exists a least $w' \in N$ such that $w' \to (3c+1)_c^{2c+1}$. We have already chosen such a number for $\mathcal{M}$, namely w, and we have chosen it so that it is minimal with this property in $\mathcal{M}$.

Can it be that $w' < w$? Since $\mathcal{N} \subseteq \mathcal{M}$ (and hence $w' \in M$), this seems to contradict the minimality of w in $\mathcal{M}$. This is indeed so, but again things are a little more complicated because we are working with non-standard models. The formalized version of the finite Ramsey theorem works via codes, in the sense that

for every p that *codes* a function $[w']^{2c+1} \to c$, there exists a *code* q for a homogeneous set of size $\geq 3c+1$.

Since c is non-standard, there will be a lot of possible functions $[w']^{2c+1} \to c$ (uncountably many), and not all of them might be coded in $\mathcal{N}$. But again due to the simple (i.e., easily definable) nature of codes, every such set and function coded in $\mathcal{M}$ is also coded in $\mathcal{N}$. This means that $w' < w$ would indeed contradict the minimality of w in $\mathcal{M}$. Therefore $w \leq w'$, and hence $w \in N$.

Now, since $\mathcal{N} \vDash \mathsf{PA} + (*)$ *and* $w \in N$, we can execute the proof of Principle $(*)$ (Lemma 4.52) inside $\mathcal{N}$ and obtain W' such that every regressive function defined on $[W']^n$ has a min-homogeneous set H of cardinality w with $\min H \geq c$. By the same argument as above, this W' would also work for $\mathcal{M}$. But $W \notin N$ and hence $W' < W$, contradicting the minimality of W in $\mathcal{M}$. Therefore, $\mathcal{N} \nvDash (*)$. This completes the proof.

We finish with a brief summary of the argument: *W is the minimal number that allows us to find min-homogeneous functions of a certain size inside $\mathcal{M}$ (as $(*)$ holds in $\mathcal{M}$). The min-homogeneous functions, in turn, allow us to find diagonal indiscernibles, which in turn yield a model $\mathcal{N}$ of PA inside $\mathcal{M}$ which does not contain W. Because of the way codes and Gödel numbers subsist between models of PA, $(*)$ cannot hold in $\mathcal{N}$, as it does not contain the "witness" W.*

4.8. More incompleteness

We mentioned at the end of Section 3.5 that the diagonal Paris-Harrington function

$$F(x) = PH(x+1, x, x)$$

will eventually dominate every function Φ_α with $\alpha < \varepsilon_0$. Ketonen and Solovay's proof [40] of this fact uses only elementary combinatorics. There is also a metamathematical argument for this, and it is implicit in the Paris-and-Harrington theorem.

Let us call a function $f : \mathbb{N} \to \mathbb{N}$ **provably total in** PA if there exists an $\mathcal{L}_A$-formula $\varphi(x, y)$ such that

(i) φ is of the form $\exists z\, \psi(x, y, z)$ where ψ is Δ_0;

(ii) for all $m, n \in \mathbb{N}$, PA $\vdash \varphi(\underline{m}, \underline{n})$ if and only if $f(m) = n$; and

(iii) PA $\vdash \forall x \exists y\, \varphi(x, y)$.

Intuitively, f is provably total if there exists a reasonably simple formula as in (i) that correctly defines f in PA, as in (ii), and PA can prove that φ defines a total function, as in (iii).

Wainer [71] showed that the growth of provably total functions is bounded by Φ_{ε_0}.

Theorem 4.56. *If $f : \mathbb{N} \to \mathbb{N}$ is provably total in* PA, *then f is eventually dominated by some Φ_α with $\alpha < \varepsilon_0$. Conversely, for every $\alpha < \varepsilon_0$, Φ_α is provably total.*

The proof of the Paris-Harrington theorem can be adapted to show that $F(x)$ eventually dominates every function that is provably total in PA. The key idea is that if g is provably total, $g(x) \geq F(x)$ infinitely often, and one can use a compactness argument to show that there exists a non-standard model $\mathcal{M}$ of PA with a non-standard element $a \in M$ such that $F(a) \leq g(a)$. One can then use the method of diagonal indiscernibles to construct a cut in $\mathcal{M}$ that is bounded by $g(a)$. This means that $g(a)$ cannot be in the induced model $\mathcal{N}$ of PA, and therefore $\mathcal{N} \nvDash \exists y\, (g(a) = y)$. This contradicts the assumption that g is provably total.

Together with Wainer's result, this yields the lower bound on the growth of the diagonal Paris-Harrington function.

Theorem 4.57. $F(x) = PH(x+1, x, x)$ eventually dominates every function Φ_α with $\alpha < \varepsilon_0$.

On the other hand, if one combines Ketonen and Solovay's combinatorial proof of Theorem 4.57 with Wainer's analysis of the provably total functions, one obtains an alternative proof of the Paris-Harrington theorem.

The second Gödel incompleteness theorem. The result by Paris and Harrington gives us an example of a statement that is true in ℕ but not provable in PA. An essential ingredient in the proof is the fact that the fast Ramsey theorem allows one to construct models of PA, and this construction can be formalized in PA.

This touches directly on Gödel's *second incompleteness theorem*. Combining Gödelization for formulas and sequences of natural numbers, we can devise a Gödel numbering for sequences of formulas. Given a Gödel number $x = \ulcorner \sigma \urcorner$ and another Gödel number $y = \ulcorner \langle \varphi_1, \ldots, \varphi_n \rangle \urcorner$, we can ask whether y represents a proof of x. Since we can effectively check this, the relation

$$\text{Proof}(x,y) :\Leftrightarrow (y \text{ codes a proof in PA for the formula coded by } x)$$

is decidable. That is, there exists a Turing machine M such that

$$M(\langle x, y \rangle) = \begin{cases} 1 & \text{if Proof}(x,y) \text{ holds}, \\ 0 & \text{if Proof}(x,y) \text{ does not hold}. \end{cases}$$

Let e be a Gödel number for this Turing machine. Using the predicate Ψ from Theorem 4.37, we define

$$\text{Provable}(x) \equiv \exists y\, \Psi(\underline{e}, \langle x, y \rangle, 1).$$

Provable(x) is an $\mathcal{L}_{\mathcal{A}}$-formula that expresses that the formula with Gödel number x is provable in PA. We can use the predicate Provable to make statements about PA "inside" PA. For example, let us define the sentence

$$\text{Con}_{\mathsf{PA}} \equiv \neg \text{Provable}(\ulcorner 0 \neq 0 \urcorner).$$

This is an $\mathcal{L}_{\mathcal{A}}$-formula asserting that PA cannot prove $0 \neq 0$, in other words, asserting that PA is consistent. A special case of Gödel's second incompleteness theorem states the following.

4.8. More incompleteness

Theorem 4.58 (Gödel [**19**]). *If* PA *is consistent,* Con_{PA} *is not provable in* PA.

Corollary 4.8 established that a theory is consistent if and only if it has a model. And clearly (or not so clearly?), $\mathbb{N}$ is a model of PA. Therefore, PA must be consistent. The problem with this argument is that it cannot be formalized in PA (this is what Gödel's result says). If we assume the fast Ramsey theorem, however, Paris and Harrington showed that we can use it to construct models of PA *and* this proof can be formalized in PA.

Theorem 4.59 (Paris and Harrington [**47**]). *The fast Ramsey theorem implies* Con_{PA}.

The fast Ramsey theorem turns out to be a powerful metamathematical principle. While it looks like a finitary statement about natural numbers, its true nature seems to transcend the finite world. It now seems much closer to infinitary principles such as compactness.

These results offer a first glimpse into what has become a very active area in mathematical logic: *reverse mathematics*. The basic question is to take a mathematical theorem and ask which foundational principles or axioms it implies. We *reverse* the usual nature of mathematical inquiry: Instead of proving theorems from the axioms, we try to prove the axioms from the theorems. Ramsey theory has proved to be a rich and important source for this endeavor, and the Paris-Harrington theorem is just one aspect of it. We have seen that compactness can be used to prove many results. Now one can ask, for example, whether the infinite Ramsey theorem in turn implies the compactness principle. We cannot answer these questions here, but instead refer the interested reader to the wonderful book by Hirschfeldt [**34**] or the classic by Simpson [**61**].

Incompleteness in set theory. Gödel showed that the second incompleteness theorem applies not only to PA but to any consistent computable extension of PA. That is, no such axiom system can prove its own consistency. Moreover, if one can define a version of PA inside another system (one says that the other system *interprets* PA), the second incompleteness theorem holds for these systems, too. One important example is **Zermelo-Fraenkel set theory** with the axiom of choice, ZFC.

Just as PA collects basic statements about natural numbers, ZFC consists of various statements about sets. For example, one axiom asserts that if X and Y are sets, so is $\{X, Y\}$. Another axiom asserts the existence of the power set $\mathcal{P}(X)$ for any set X. You can find the complete list in any book on set theory (such as [**35**]). ZFC is a powerful axiom system. Most mathematical objects and theories (from analysis to group theory to algebraic topology) can be formalized in it. ZFC interprets PA and one can also formalize the proof that $\mathbb{N}$ is a model of PA in ZFC, which means that ZFC proves the consistency of PA. However, it also means that ZFC cannot prove its own consistency, in the sense formulated above. One would have to resort to an even stronger axiom system to prove the consistency of ZFC. The stronger system, if consistent, in turn cannot prove its own consistency, and so on.

There is something similar to a standard model of PA in ZFC: the **von Neumann universe** V. It is a cumulative hierarchy of sets, built from the empty set by iterating the power set operation and taking unions: For ordinals α and λ, we define

$$V_0 = \varnothing,$$
$$V_{\alpha+1} = \mathcal{P}(V_\alpha), \text{ and}$$
$$V_\lambda = \bigcup_{\beta < \lambda} V_\beta \text{ if } \lambda \text{ is a limit ordinal.}$$

The proper class

$$V = \bigcup_{\alpha \in \mathrm{Ord}} V_\alpha$$

satisfies all the axioms of ZFC, but by the second incompleteness theorem, the proof of this cannot be formalized in ZFC.

We ended Chapter 2 with the introduction of Ramsey cardinals. We did not answer the question of whether Ramsey cardinals exist back then. We will not be able to do this now either, but we can at least explain a bit more about why this question is difficult to answer.

4.8. More incompleteness

In Theorem 2.46, we showed that every Ramsey cardinal is inaccessible. Inaccessible cardinals are important stages in the von Neumann hierarchy.

Theorem 4.60. *The following is provable in* ZFC*:*
κ *inaccessible* $\Rightarrow$ *all axioms of* ZFC *hold in* V_κ.

Therefore, if ZFC is consistent and could prove the existence of an inaccessible cardinal, ZFC could prove the existence of a model of ZFC, which by the completeness theorem implies the consistency of ZFC. Thus, ZFC could prove its own consistency, which is impossible by the second incompleteness theorem.

It follows that the existence of inaccessible cardinals, and therefore that of Ramsey cardinals, cannot be proved in ZFC. Even worse, if ZFC is consistent, we cannot even show that the existence of inaccessible cardinals is *consistent* with ZFC (another consequence of the second incompleteness theorem; see [**35**]).

There seems to be a murky abyss lurking at the bottom of mathematics. While in many ways we cannot hope to reach solid ground, mathematicians have built impressive ladders that let us explore the depths of this abyss and marvel at the limits and at the power of mathematical reasoning at the same time. Ramsey theory is one of those ladders.

Bibliography

[1] W. Ackermann, *Zum Hilbertschen Aufbau der reellen Zahlen* (German), Math. Ann. **99** (1928), no. 1, 118–133, DOI 10.1007/BF01459088. MR1512441

[2] N. Alon and J. H. Spencer, *The probabilistic method*, 4th ed., Wiley Series in Discrete Mathematics and Optimization, John Wiley & Sons, Inc., Hoboken, NJ, 2016. MR3524748

[3] V. Angeltveit and B. D. McKay, $R(5,5) \leq 48$, arXiv:1703.08768, 2017.

[4] R. A. Brualdi, *Introductory combinatorics*, 5th ed., Pearson Prentice Hall, Upper Saddle River, NJ, 2010. MR2655770

[5] P. L. Butzer, M. Jansen, and H. Zilles, *Johann Peter Gustav Lejeune Dirichlet (1805–1859): Genealogie und Werdegang* (German), Dürerner Geschichtsblätter **71** (1982), 31–56. MR690659

[6] A. Church, *A note on the Entscheidungsproblem.*, Journal of Symbolic Logic **1** (1936), 40–41.

[7] A. Church, *An unsolvable problem of elementary number theory*, Amer. J. Math. **58** (1936), no. 2, 345–363, DOI 10.2307/2371045. MR1507159

[8] P. J. Cohen, *The independence of the continuum hypothesis*, Proc. Nat. Acad. Sci. U.S.A. **50** (1963), 1143–1148. MR0157890

[9] P. J. Cohen, *The independence of the continuum hypothesis. II*, Proc. Nat. Acad. Sci. U.S.A. **51** (1964), 105–110. MR0159745

[10] D. Conlon, *A new upper bound for diagonal Ramsey numbers*, Ann. of Math. (2) **170** (2009), no. 2, 941–960, DOI 10.4007/annals.2009.170.941. MR2552114

[11] R. Dedekind, *Was sind und was sollen die Zahlen?* (German), 8te unveränderte Aufl, Friedr. Vieweg & Sohn, Braunschweig, 1960. MR0106846

[12] R. Diestel, *Graph theory*, 5th ed., Graduate Texts in Mathematics, vol. 173, Springer, Berlin, 2017. MR3644391

[13] H. B. Enderton, *A mathematical introduction to logic*, 2nd ed., Harcourt/Academic Press, Burlington, MA, 2001. MR1801397

[14] P. ErdH øs, *Some remarks on the theory of graphs*, Bull. Amer. Math. Soc. **53** (1947), 292–294, DOI 10.1090/S0002-9904-1947-08785-1. MR0019711

[15] P. Erdős and R. Rado, *A problem on ordered sets*, J. London Math. Soc. **28** (1953), 426–438, DOI 10.1112/jlms/s1-28.4.426. MR0058687

[16] L. Euler, *Solutio problematis ad geometriam situs pertinentis*, Commentarii Academiae Scientiarum Imperialis Petropolitanae **8** (1736), 128–140.

[17] H. Furstenberg, *Ergodic behavior of diagonal measures and a theorem of Szemerédi on arithmetic progressions*, J. Analyse Math. **31** (1977), 204–256, DOI 10.1007/BF02813304. MR0498471

[18] C. F. Gauss, *Disquisitiones arithmeticae*, Springer-Verlag, New York, 1986. Translated and with a preface by Arthur A. Clarke; Revised by William C. Waterhouse, Cornelius Greither and A. W. Grootendorst and with a preface by Waterhouse. MR837656

[19] K. Gödel, *Über formal unentscheidbare Sätze der Principia Mathematica und verwandter Systeme I* (German), Monatsh. Math. Phys. **38** (1931), no. 1, 173–198, DOI 10.1007/BF01700692. MR1549910

[20] K. Gödel, *The Consistency of the Continuum Hypothesis*, Annals of Mathematics Studies, no. 3, Princeton University Press, Princeton, NJ, 1940. MR0002514

[21] G. Gonthier, A. Asperti, J. Avigad, et al., *A machine-checked proof of the odd order theorem*, Interactive theorem proving, Lecture Notes in Comput. Sci., vol. 7998, Springer, Heidelberg, 2013, pp. 163–179, DOI 10.1007/978-3-642-39634-2_14. MR3111271

[22] W. T. Gowers, *A new proof of Szemerédi's theorem*, Geom. Funct. Anal. **11** (2001), no. 3, 465–588, DOI 10.1007/s00039-001-0332-9. MR1844079

[23] R. L. Graham and B. L. Rothschild, *Ramsey theory* (1980), ix+174. Wiley-Interscience Series in Discrete Mathematics; A Wiley-Interscience Publication. MR591457

[24] R. L. Graham, B. L. Rothschild, and J. H. Spencer, *Ramsey theory*, 2nd ed., Wiley-Interscience Series in Discrete Mathematics and Optimization, John Wiley & Sons, Inc., New York, 1990. A Wiley-Interscience Publication. MR1044995

[25] R. L. Graham and J. H. Spencer, *Ramsey theory*, Scientific American **263** (1990), no. 1, 112–117.

[26] B. Green and T. Tao, *The primes contain arbitrarily long arithmetic progressions*, Ann. of Math. (2) **167** (2008), no. 2, 481–547, DOI 10.4007/annals.2008.167.481. MR2415379

[27] R. E. Greenwood and A. M. Gleason, *Combinatorial relations and chromatic graphs*, Canad. J. Math. **7** (1955), 1–7, DOI 10.4153/CJM-1955-001-4. MR0067467

[28] A. Grzegorczyk, *Some classes of recursive functions*, Rozprawy Mat. **4** (1953), 46. MR0060426

[29] A. W. Hales and R. I. Jewett, *Regularity and positional games*, Trans. Amer. Math. Soc. **106** (1963), 222–229, DOI 10.2307/1993764. MR0143712

[30] T. Hales, M. Adams, G. Bauer, T. D. Dang, J. Harrison, Le Truong Hoang, C. Kaliszyk, V. Magron, S. McLaughlin, T. T. Nguyen, Q. T. Nguyen, T. Nipkow, S. Obua, J. Pleso, J. Rute, A. Solovyev, T. H. A. Ta, N. T. Tran, T. D. Trieu, J. Urban, K. Vu, and R. Zumkeller, *A formal proof of the Kepler conjecture*, Forum Math. Pi **5** (2017), e2, 29, DOI 10.1017/fmp.2017.1. MR3659768

[31] G. H. Hardy and E. M. Wright, *An introduction to the theory of numbers*, Oxford, at the Clarendon Press, 1954. 3rd ed. MR0067125

[32] D. Hilbert and W. Ackermann, *Grundzüge der theoretischen Logik*, Springer-Verlag, Berlin, 1928.

[33] N. Hindman and E. Tressler, *The first nontrivial Hales-Jewett number is four*, Ars Combin. **113** (2014), 385–390. MR3186481

Bibliography

[34] D. R. Hirschfeldt, *Slicing the truth. On the computable and reverse mathematics of combinatorial principles*, Lecture Notes Series. Institute for Mathematical Sciences. National University of Singapore, vol. 28, World Scientific Publishing Co. Pte. Ltd., Hackensack, NJ, 2015. Edited and with a foreword by Chitat Chong, Qi Feng, Theodore A. Slaman, W. Hugh Woodin and Yue Yang. MR3244278

[35] T. J. Jech, *Set theory*, Springer Monographs in Mathematics, Springer-Verlag, Berlin, 2003. The third millennium edition, revised and expanded. MR1940513

[36] T. J. Jech, *The axiom of choice*, North-Holland Publishing Co., Amsterdam-London; Amercan Elsevier Publishing Co., Inc., New York, 1973. Studies in Logic and the Foundations of Mathematics, Vol. 75. MR0396271

[37] C. G. Jockusch Jr., *Ramsey's theorem and recursion theory*, J. Symbolic Logic **37** (1972), 268–280, DOI 10.2307/2272972. MR0376319

[38] A. Kanamori and K. McAloon, *On Gödel incompleteness and finite combinatorics*, Ann. Pure Appl. Logic **33** (1987), no. 1, 23–41, DOI 10.1016/0168-0072(87)90074-1. MR870685

[39] R. Kaye, *Models of Peano arithmetic*, Oxford Logic Guides, vol. 15, The Clarendon Press, Oxford University Press, New York, 1991. Oxford Science Publications. MR1098499

[40] J. Ketonen and R. Solovay, *Rapidly growing Ramsey functions*, Ann. of Math. (2) **113** (1981), no. 2, 267–314, DOI 10.2307/2006985. MR607894

[41] A. Y. Khinchin, *Three pearls of number theory*, Graylock Press, Rochester, NY, 1952. MR0046372

[42] D. E. Knuth, *Mathematics and computer science: coping with finiteness*, Science **194** (1976), no. 4271, 1235–1242, DOI 10.1126/science.194.4271.1235. MR534161

[43] B. M. Landman and A. Robertson, *Ramsey theory on the integers*, 2nd ed., Student Mathematical Library, vol. 73, American Mathematical Society, Providence, RI, 2014. MR3243507

[44] D. Marker, *Model theory: an introduction*, Graduate Texts in Mathematics, vol. 217, Springer-Verlag, New York, 2002. MR1924282

[45] B. D. McKay and S. P. Radziszowski, $R(4,5) = 25$, J. Graph Theory **19** (1995), no. 3, 309–322, DOI 10.1002/jgt.3190190304. MR1324481

[46] J. Nešetřil, *Ramsey theory*, Handbook of combinatorics, Vol. 1, 2, Elsevier Sci. B. V., Amsterdam, 1995, pp. 1331–1403. MR1373681

[47] J. Paris and L. Harrington, *A mathematical incompleteness in Peano arithmetic*, Handbook of mathematical logic, Stud. Logic Found. Math., vol. 90, North-Holland, Amsterdam, 1977, pp. 1133–1142. MR3727432

[48] G. Peano, *Arithmetices principia: nova methodo*, Fratres Bocca, 1889.

[49] G. Peano, *Sul concetto di numero*, Rivista di Matematica **1** (1891), 256–267.

[50] R. Péter, *Über die mehrfache Rekursion* (German), Math. Ann. **113** (1937), no. 1, 489–527, DOI 10.1007/BF01571648. MR1513105

[51] C. C. Pugh, *Real mathematical analysis*, 2nd ed., Undergraduate Texts in Mathematics, Springer, Cham, 2015. MR3380933

[52] S. Radziszowski, *Small Ramsey numbers*, Electron. J. Comb., http://www.combinatorics.org/files/Surveys/ds1/ds1v15-2017.pdf, 2017.

[53] F. P. Ramsey, *On a problem of formal logic*, Proc. London Math. Soc. (2) **30** (1929), no. 4, 264–286, DOI 10.1112/plms/s2-30.1.264. MR1576401

[54] W. Rautenberg, *A concise introduction to mathematical logic*, Based on the second (2002) German edition, Universitext, Springer, New York, 2006. With a foreword by Lev Beklemishev. MR2218537

[55] T. Ridge, *Hol/library/ramsey.thy*.

Bibliography

[56] H. E. Rose, *Subrecursion: functions and hierarchies*, Oxford Logic Guides, vol. 9, The Clarendon Press, Oxford University Press, New York, 1984. MR752696

[57] P. Rothmaler, *Introduction to model theory*, Algebra, Logic and Applications, vol. 15, Gordon and Breach Science Publishers, Amsterdam, 2000. Prepared by Frank Reitmaier; Translated and revised from the 1995 German original by the author. MR1800596

[58] S. Shelah, *Primitive recursive bounds for van der Waerden numbers*, J. Amer. Math. Soc. **1** (1988), no. 3, 683–697, DOI 10.2307/1990952. MR929498

[59] L. Shi, *Upper bounds for Ramsey numbers*, Discrete Math. **270** (2003), no. 1-3, 251–265, DOI 10.1016/S0012-365X(02)00837-3. MR1997902

[60] J. R. Shoenfield, *Mathematical logic*, Association for Symbolic Logic, Urbana, IL; A K Peters, Ltd., Natick, MA, 2001. Reprint of the 1973 second printing. MR1809685

[61] S. G. Simpson, *Subsystems of second order arithmetic*, 2nd ed., Perspectives in Logic, Cambridge University Press, Cambridge; Association for Symbolic Logic, Poughkeepsie, NY, 2009. MR2517689

[62] C. Smoryński, *Logical number theory. I. An introduction*, Universitext, Springer-Verlag, Berlin, 1991. MR1106853

[63] R. I. Soare, *Turing computability: theory and applications*, Theory and Applications of Computability, Springer-Verlag, Berlin, 2016. MR3496974

[64] E. Szemerédi, *On sets of integers containing no k elements in arithmetic progression*, Acta Arith. **27** (1975), 199–245, DOI 10.4064/aa-27-1-199-245. Collection of articles in memory of Juriĭ Vladimirovič Linnik. MR0369312

[65] P. Turán, *Eine Extremalaufgabe aus der Graphentheorie* (Hungarian, with German summary), Mat. Fiz. Lapok **48** (1941), 436–452. MR0018405

[66] A. M. Turing, *On computable numbers, with an application to the Entscheidungsproblem*, Proc. London Math. Soc. (2) **42** (1936), no. 3, 230–265, DOI 10.1112/plms/s2-42.1.230. MR1577030

[67] S. M. Ulam, *Adventures of a mathematician*, University of California Press, 1991.

[68] B. L. van der Waerden, *Beweis einer Baudetschen Vermutung*, (German), Nieuw Arch. Wiskd., II. Ser. **15** (1927), 212–216.

[69] B. L. van der Waerden, *Wie der Beweis der Vermutung von Baudet gefunden wurde* (German), Abh. Math. Sem. Univ. Hamburg **28** (1965), 6–15, DOI 10.1007/BF02993133. MR0175875

[70] B. L. van der Waerden, *How the proof of Baudet's conjecture was found*, Studies in Pure Mathematics (Presented to Richard Rado), Academic Press, London, 1971, pp. 251–260. MR0270881

[71] S. S. Wainer, *A classification of the ordinal recursive functions*, Arch. Math. Logik Grundlagenforsch. **13** (1970), 136–153, DOI 10.1007/BF01973619. MR0294134

[72] S. Weinberger, *Computers, rigidity, and moduli. The large-scale fractal geometry of Riemannian moduli space*, M. B. Porter Lectures, Princeton University Press, Princeton, NJ, 2005. MR2109177

Notation

Symbol	Meaning	Page
$[n]$	the set of integers from 1 to n	1
$\|S\|$	cardinality of a set S	2
$[S]^p$	the set of p-element subsets of S	2
$[n]^p$	the set of p-element subsets of $[n]$	2
$N \to (k)^p_r$	every r-coloring of $[S]^p$ with $\|S\| \geq N$ has a k-element monochromatic subset	2
K_n	complete graph on n vertices	9
$K_{n,m}$	complete bipartite graph of order n, m	11
$R(n)$	nth Ramsey number	21
$R(m,n)$	generalized Ramsey number	22
ω	first infinite ordinal	56
Ord	class of all ordinal numbers	57
ε_0	least ordinal such that $\omega^{\varepsilon_0} = \varepsilon_0$	59
$\mathcal{P}(S)$	power set of S, $\mathcal{P}(S) = \{A : A \subseteq S\}$	68

Notation

Symbol	Meaning	Page
κ^+	the least cardinal greater than κ	68
2^κ	the cardinality of the power set of a set of cardinality κ	69
$\aleph_0$	the first infinite cardinal, $\aleph_0 = |\mathbb{N}|$	69
$W(k,r)$	van der Waerden number for k-APs and r-coloriing	86
$U(k,r)$	upper bound for $W(k,r)$, extracted from the proof	96
$\varphi(x,y,z)$	Ackermann function	102
$\varphi_n(x,y)$	nth level Ackermann function	103
C_t^n	combinatorial cube of dimension n with side length t	113
$HJ(t,r)$	Hales-Jewett number for side length t symbols and r colors	115
$PH(m,p,r)$	Paris-Harrington number	123
$\mathcal{L}_\mathcal{A}$	language of arithmetic, $\{S, +, \cdot, 0\}$	133
PA	Peano arithmetic	135
$\mathcal{A} \vDash \sigma$	σ holds in structure $\mathcal{A}$	138
$A \vdash \sigma$	A proves σ	141
Th($\mathbb{N}$)	theory of the natural numbers	160
$\ulcorner \varphi \urcorner$	Gödel number of formula φ	164
ZFC	Zermelo-Fraenkel set theory with the axiom of choice	196

Index

Ackermann function, 102, 105, 106, 108, 110
alephs, 69
arithmetic progression, 85
arrow notation, 2
axiom of choice, 63, 67

Banach-Tarksi paradox, 64
binary string, 46
bounded μ-operator, 107
Burali-Forti paradox, 57

cardinal, 67
  inaccessible, 83, 197
  limit, 81
  Ramsey, 83, 197
  regular, 82
  singular, 82
  strong limit, 81
cardinal arithmetic, 68
cardinality (of a set), 67
Church-Turing thesis, 163
closed, 51
cofinality, 82, 111
coloring, 2
  edge, 14
combinatorial
  s-space, 118
combinatorial line, 114
compactness, xi, 49, 86, 123, 144
  sequentially compact, 52
  topological, 52
complete (proof system), 142, 160
complete (theory), 159
computable, 162
continuum hypothesis, 69
  generalized, 70, 81
cut, 149

definable, 149, 155
degree (graphs), 6
dominating function, 99
eventually, 99

equivalence relation, 8, 66

free variable, 134
fundamental sequence, 111

Gödel β-function, 155, 190
Gödel number, 164
Goldbach conjecture, 129
graph, 5
  k-partite, 11
  bipartite, 10, 31
  clique, 10
  complement, 6
  complete, 9
  complete bipartite, 11
  connected, 9
  cycle, 8

Index

hypergraph, 34
independent, 10
induced subgraph, 7
isomorphic, 5
order, 5
Paley graph, 26
path, 8
subgraph, 7
tree, 12
Turán, 32
Grzegorczyk hierarchy, 109, 123, 124

Hales-Jewett numbers, 115
halting problem, 166
homogeneous, 41
min-homogeneous, 182, 192

inconsistent, 143
indiscernibles, 171
diagonal, 173
order, 172
induction scheme (axiom), 135

Knuth arrow notation, 101, 103

least number principle (LNP), 136, 181

metric, 50
discrete, 51
Euclidean, 51
path, 53

neighborhood, 51
non-standard number, 147

open, 51
order
linear, 45
order type, 62
partial, 13, 45
well-ordering, 59
ordinal, 55
addition, 57
exponentiation, 59
limit, 57
multiplication, 58
successor, 56
overspill, 151

pairing function, 65
Peano arithmetic, 135, 139, 145, 152, 161, 168, 171, 188
non-standard model, 145, 171, 189
standard model, 139
pigeonhole principle, 17, 18, 34, 39, 90, 118, 186
infinite, 41, 42, 72
power set, 34
primitive recursive, 106, 126, 156, 163
principle (∗), 185, 187, 188
probabilistic method, 29
proof, 136
system, 137
provably total, 193

Ramsey number, 21
generalized, 22
regressive, 185
relatively large, 123
reverse mathematics, 195

satisfiable, 143
sentence (formula), 134
Shelah
s-space, 119
line, 117
point, 118
soundness (proof system), 142
star word, 115
structure, 138

tetration, 100
theorem
Bolzano-Weierstrass, 52
Cantor cardinality theorem, 68
Cantor normal form, 111
Cantor-Schröder-Bernstein, 65
Chinese remainder theorem, 154
compactness (logic), 144
Erdős-Rado, 76
fast Ramsey, 170, 185, 187, 189, 195
finite Ramsey, 34, 43, 48, 123, 152, 183, 192
first Gödel incompleteness, 168
Gödel incompleteness, xi

Index

Gödel completeness, 142
Greenwood and Gleason bound, 22
Hales-Jewett, 113
Heine-Borel, 52
infinite Ramsey, 41, 52, 71, 124, 172
König's lemma, 48, 49, 54, 124
Paris-Harrington, xi, 170, 193
Ramsey (for graphs), 16, 37
Schur, 21
second Gödel incompleteness, 195, 197
Szemerédi, 98
Turán, 31, 98
unsolvability Entscheidungsproblem, 167
unsolvability of halting problem, 166
van der Waerden, ix, 86, 115, 159
tree, 12, 46
 binary, 46
 finitely branching, 47
 infinite path, 47
Turing machine, 162

van der Waerden number, 86

Wainer hierarchy, 113
well-ordering principle, 63

Zermelo-Fraenkel set theory with choice (ZF), 69
Zermelo-Fraenkel set theory with choice (ZFC), 196

SELECTED PUBLISHED TITLES IN THIS SERIES

87 **Matthew Katz and Jan Reimann,** An Introduction to Ramsey Theory, 2018
86 **Peter Frankl and Norihide Tokushige,** Extremal Problems for Finite Sets, 2018
85 **Joel H. Shapiro,** Volterra Adventures, 2018
84 **Paul Pollack,** A Conversational Introduction to Algebraic Number Theory, 2017
83 **Thomas R. Shemanske,** Modern Cryptography and Elliptic Curves, 2017
82 **A. R. Wadsworth,** Problems in Abstract Algebra, 2017
81 **Vaughn Climenhaga and Anatole Katok,** From Groups to Geometry and Back, 2017
80 **Matt DeVos and Deborah A. Kent,** Game Theory, 2016
79 **Kristopher Tapp,** Matrix Groups for Undergraduates, Second Edition, 2016
78 **Gail S. Nelson,** A User-Friendly Introduction to Lebesgue Measure and Integration, 2015
77 **Wolfgang Kühnel,** Differential Geometry: Curves — Surfaces — Manifolds, Third Edition, 2015
76 **John Roe,** Winding Around, 2015
75 **Ida Kantor, Jiří Matoušek, and Robert Šámal,** Mathematics++, 2015
74 **Mohamed Elhamdadi and Sam Nelson,** Quandles, 2015
73 **Bruce M. Landman and Aaron Robertson,** Ramsey Theory on the Integers, Second Edition, 2014
72 **Mark Kot,** A First Course in the Calculus of Variations, 2014
71 **Joel Spencer,** Asymptopia, 2014
70 **Lasse Rempe-Gillen and Rebecca Waldecker,** Primality Testing for Beginners, 2014
69 **Mark Levi,** Classical Mechanics with Calculus of Variations and Optimal Control, 2014
68 **Samuel S. Wagstaff, Jr.,** The Joy of Factoring, 2013
67 **Emily H. Moore and Harriet S. Pollatsek,** Difference Sets, 2013
66 **Thomas Garrity, Richard Belshoff, Lynette Boos, Ryan Brown, Carl Lienert, David Murphy, Junalyn Navarra-Madsen, Pedro Poitevin, Shawn Robinson, Brian Snyder, and Caryn Werner,** Algebraic Geometry, 2013
65 **Victor H. Moll,** Numbers and Functions, 2012

For a complete list of titles in this series, visit the
AMS Bookstore at **www.ams.org/bookstore/stmlseries/**.